SpringerBriefs in Computer Science

Series Editors
Stan Zdonik
Peng Ning
Shashi Shekhar
Jonathan Katz
Xindong Wu
Lakhmi C. Jain
David Padua
Xuemin Shen
Borko Furht
VS Subrahmanian

For further volumes:
http://www.springer.com/series/10028

Mohammad Mehedi Hassan • Eui-Nam Huh

Dynamic Cloud Collaboration Platform

A Market-Oriented Approach

 Springer

Mohammad Mehedi Hassan
College of Computer and Information
    Sciences
Chair of Pervasive and Mobile Computing
King Saud University
Riyadh, Kingdom of Saudi Arabia

Eui-Nam Huh
College of Electronics and Information
Department of Computer Engineering
Kyung Hee University, Global Campus
Gyeonggi-do, South Korea

ISSN 2191-5768   ISSN 2191-5776 (electronic)
ISBN 978-1-4614-5145-7 ISBN 978-1-4614-5146-4 (eBook)
DOI 10.1007/978-1-4614-5146-4
Springer New York Heidelberg Dordrecht London

Library of Congress Control Number: 2012945400

Printed on acid-free paper

Springer is part of Springer Science+Business Media (www.springer.com)

Preface

The purpose of this book is to develop solutions to enable dynamic collaboration of multiple cloud service providers (CPs) to ensure quality of service (QoS)-constrained service delivery, by combating against resource over-provisioning, service level agreement (SLA) violation, and excessive operational cost for a cloud provider. Throughout the book, we introduce landmark achievements towards realizing this aim.

Present trends in cloud provider capabilities give rise to the interest in federating or collaborating clouds, thus allowing providers to revel on increased scale and reach more than that is achievable individually. Current research efforts in this context mainly focus on building supply chain collaboration (SCC) models in which CPs leverage cloud services from other CPs for seamless provisioning. Nevertheless, in the near future, we can expect that hundreds of CPs will compete to offer services and thousands of users also compete to receive the services to run their complex heterogeneous applications on cloud computing environment. In this open federation scenario, existing collaboration models (i.e., SCC) are not applicable. In fact, while clouds are typically heterogeneous and dynamic, the existing collaboration models are designed for static environments where a priori agreements among the parties are needed to establish the federation.

To move beyond these shortcomings, the book establishes the basis for developing advanced and efficient collaborative cloud service solutions that are scalable, of high performance, and cost effective. We term the technology for interconnection and inter-operation of CPs in open cloud federation as Dynamic Cloud Collaboration (DCC) in which various CPs (smaller, medium, and large) of complementary service requirements will collaborate dynamically to gain economies of scale and enlargements of their capabilities to meet QoS requirements of consumers. In this context, this book addresses four key issues—when to collaborate (triggering circumstances), whom to collaborate with (suitable partners), how to collaborate (architectural model), and how to demonstrate collaboration applicability (simulation study). It also provides candidate solutions to them, which are effective in real environments.

Contents

Chapter 1
Overview of Cloud Computing and Motivation of the Work

Abstract This chapter provides an overview of cloud computing, which can help the readers to understand the basics of cloud computing technology as well as motivation of the current work towards enabling dynamic cloud federation platform.

1.1 Introduction

During the recent years, cloud computing is gaining more attention from academic as well as commercial world. The business potential of cloud computing is recognized by several market research firms including IDC, which reports that worldwide spending on cloud services will make up $17.4 billion worth of IT purchases and be a $44.2 billion market by 2013 [47]. Providers such as Amazon [3], Google [4], Salesforce [76], IBM [54], Microsoft [5], and Sun Microsystems [83] have begun to establish new data centers for hosting cloud Computing services. Emerging cloud Computing offers a new computing model in which resources such as computing power, storage, online applications and networking infrastructures can be shared as "services" over the Internet. cloud providers (CPs) are motivated by the profits to be made by charging consumers for accessing these services. Consumers such as enterprises, IT companies, application developers and personal users are attracted by the opportunity for reducing or eliminating costs associated with "in-house" provisions of these services. Thus it enables consumers to give more focus on innovation and creating business value for their services. Also consumers access services based on their application QoS (Quality of Service) requirements without regard to where the services are hosted [16]. Some examples of emerging cloud applications include social networking, gaming portals, business applications, media content delivery, and scientific workflows.

M.M. Hassan and E.-N. Huh, *Dynamic Cloud Collaboration Platform: A Market-Oriented Approach*, SpringerBriefs in Computer Science, DOI 10.1007/978-1-4614-5146-4_1, © The Author(s) 2013

1.1.1 Cloud Computing: Definitions

Cloud computing is a new subject at both technological and commercial level, therefore various definitions can be found, focusing on different characteristics of cloud Computing technology, services, and platform [46]. IBM takes a technical stance and defines cloud computing as follows:

> A cloud is a pool of virtualized resources that hosts a variety of workloads, allows for a quick scale-out and deployment, provision of virtual machines to physical machines, supports redundancy and self-recovery and could also be monitored and rebalanced in real time [11].

A scientific definition is proposed by the GRIDS Lab at the University of Melbourne:

> A cloud is a type of parallel and distributed system consisting of a collection of interconnected and virtualized computers that are dynamically provisioned and presented as one or more unified computing resources based on service-level agreements (SLA) established through negotiation between the service provider and consumers [16].

Berkeley's define cloud Computing as follows:

> Cloud computing refers to both the applications delivered as services over the Internet and the hardware and systems software in the datacenters that provide those services (Software as a Service – SaaS). The datacenter hardware and software is what we will call a cloud. When a cloud is made available in a pay-as-you-go manner to the public, we call it a Public cloud; the service being sold is Utility Computing [8].

From the above definitions, the features of cloud Computing can be summarized as follows:

- It is a scalable and flexible distributed computing environment.
- It consists of a collection of interconnected and virtualized computers that are dynamically provisioned and presented as one or more unified computing resources to consumers.
- It delivers different levels of services (e.g., SaaS, PaaS, IaaS) to customers anywhere, anytime via Internet.
- It is driven by economies of scale that is the services can be dynamically configured and delivered "on-demand" at competitive costs depending on users QoS (Quality of Service) requirements.
- It provides the ability to pay for use of computing resources as needed (e.g., processors by the hour and storage by the day) and release them as needed.
- It benefits to consumers by freeing them from the low level task of setting up basic hardware (servers) and soft-ware infrastructures to run their complex applications and thus reduce the cost of "in-house" provisioning of these services.

Fig. 1.1 Layered structure of
cloud computing services

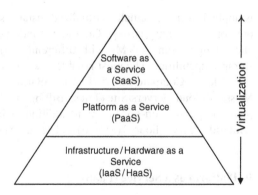

1.1.2 Cloud Types and Services

Basically clouds model can be defined by three types:

1. *Public Cloud*: Public clouds are run by third parties, and applications from different customers are likely to be mixed together on the cloud's servers, storage systems, and networks. Public clouds are most often hosted away from customer premises, and they provide a way to reduce customer risk and cost by providing a flexible, even temporary extension to enterprise infrastructure.
2. *Private Cloud*: Private clouds are built for the exclusive use of one client, providing the utmost control over data, security, and quality of service. The company owns the infrastructure and has control over how applications are deployed on it. Private clouds may be deployed in an enterprise datacenter, and they also may be deployed at a collocation facility.
3. *Hybrid Cloud*: Hybrid clouds combine both public and private cloud models. They can help to provide on-demand, externally provisioned scale. The ability to augment a private cloud with the resources of a public cloud can be used to maintain service levels in the face of rapid workload fluctuations

There are different categories of cloud services such as infrastructure, platform, software, etc. These services are delivered and consumed in real time over the Internet. Figure 1.1 below demonstrates a layered structure of current cloud Computing services, based on [99]. We discuss these services in the broader view and present some representative cloud providers of these services as follows:

A. Infrastructure as a Service (IaaS)

Infrastructure as a Service (IaaS) refers to renting raw hardware for managing its resources. In cloud Computing models those resources are virtualized and by using statistical multiplexing the "illusion of infinite resources" [8] is achieved as cloud environments can scale out. In essence, an additional layer of virtualization is applied over the actually virtualized instances of OS, which are extracted from pre-configured images (e.g. GoGrid's Windows 2003 or RedHat Linux images) and set up to run as VMs with differently adjusted RAM, CPU and hard disk storage capabilities. The most notable IaaS cloud offerings appear to be Amazon's (e.g. Elastic Compute cloud (EC2), GoGrid's cloud Hosting/Storage as well as Mosso's cloud Servers/Sites/Files. Billing models include monthly subscription plans or "pay-as-you-go" billing in addition to data transfer—"server RAM hours" (GoGrid), "server hours" (Mosso) and separate cloud storage billing.

B. Platform as a Service (PaaS)

In the context of cloud Computing, Platform as a Service (PaaS) represents an intersection between IaaS and SaaS. It is a form of SaaS that represents a platform provided to serve as the infrastructure for development and running of new applications in the cloud. Benefits include the ability to build and deploy scalable web applications without the cost and complexity of procuring server and setting them up [78]. Prominent examples of PaaS are Google AppEngine and SalesForce.com's business software development's platform—Force.com. Also Microsoft Azure Services currently supports general purpose computing by enabling applications written in a .NET language (C, VB.NET) to run in the cloud environment managed by the underlying Azure OS. The user has access to some functions that could be integrated in their application code to better take advantage of the automatic scalability properties of the distributed cloud environment (including fail over capabilities), but the user has no control whatsoever on the underlying Windows Azure OS [8].

C. Software as a Service (SaaS)

Software as a Service (SaaS), the term coined by IDC and summarized by SIIA in 2001, initially referred to the Application Service Provider (ASP) model in general, and the shift from desktop/packaged software towards web-based, outsourced solutions in particular [1]. The software is provided over a network, e.g. the Internet, but also over VPN (Virtual Private Network), and is provided to the users in a recurring fee basis. Examples of multi tenant SaaS provision from cloud managed environments are Workday.com Inc and SalesForce.com Inc. Workday offers HR and Payroll software, whereas SalesForce.com offers Customer Relationship Management (CRM) software. To highlight SaaS advantages both companies emphasize on their websites the fact their SaaS delivery models are "the opposite of ERP" and "not software". To mitigate the potential security question that arises, both companies eagerly outline the number of big enterprise customers that have entrusted them with their data.

1.1.3 Market Growth of Cloud Computing

Facing the ever larger demand of cloud computing services, various analysis institutions have mostly made optimistic predictions in the market growth of cloud Computing in the near future. Some examples are as follows:

- IDC [47] forecasted that the cloud Computing services will enjoy a growth rate of 27% CAGR in the next 4 years and reach a total market volume of $44.2 billion, accounting for 9% of overall customer spending on IT services.
- In a more aggressive prediction, Merrill Lynch [58] issued a research note said that the cloud Computing market will reach a volume of $160 billion in 2011, including $95 billion in "business and productivity applications" like office software and ERP solutions, and $65 billion in online advertising.
- In an enterprise software customer survey conducted by McKinsey and SandHill [35], 12% of the respondents claimed that they would consider using cloud Computing services.
- Gartner Inc., one of the world's leading information technology research company, has predicted the future of cloud Computing more than one time:
  - Once they said in the Gartner's Symposium ITXpo (Las Vegas) in 2008 that by 2012, 80% of Fortune 1000 companies will pay for some cloud computing service, and 30% of them will pay for cloud computing infrastructure [19].
  - A more conservative prediction from Gartner Inc. is that cloud computing services need at least 7 years to mature, so by 2015, "cloud computing will have been commoditized and will be the preferred solution for many application development projects" [45].
- As a leading provider of cloud computing service, Amazon AWS has enjoyed a quarterly growth rate of 12% during the period from 2005 to 2008 [82]; another example of how quick the cloud computing services from Amazon are expanding is that in mid 2007, the total bandwidth consumption of AWS is already more than the bandwidth consumption of Amazon's Global Websites, the websites providing the traditional ecommerce services.

1.2 Motivation and Scope of the Work

Currently cloud providers of all shapes and sizes are in a race to move as many products and services as possible to the cloud—providing managed services and software-as-a-service rather than traditional, locally-installed, software applications. However, these cloud services are proprietary in nature. They are owned and operated by individual cloud providers (public or private), each of which has created its own closed network, which is expensive to establish and maintain. Due to the proprietary nature, consumers are restricted to offerings from a single cloud provider at a time and hence cannot use multiple or collaborative cloud services at the same time. However, the rapid take-up of some services, particularly those for infrastructure, requires portability between multiple cloud infrastructures to meet different QoS targets. Thus, interoperability is also becoming an important issue for cloud services since many enterprises do not want to tie their most important applications to specific providers' remote infrastructure or platforms [9]. For example, Salesforce.com Inc's Force.com platform now enables developers to use its cloud application development platform alongside Amazon Web Services LLC's infrastructure and storage services. So to make cloud Computing truly scalable, one huge cloud that is controlled by one huge vendor running a very narrow set of applications is not feasible.

In order to support a large number of application service consumers from around the world, cloud infrastructure providers (i.e., IaaS providers) have established data centers in multiple geographical locations to provide redundancy and ensure reliability in case of site failures. However, they do not support mechanisms and policies for dynamically coordinating load distribution among different cloud-based data centers. For example, Amazon provides storage cloud service (i.e. IaaS) named Amazon EC2. It has data centers in the US (e.g., one in the East Coast and another in the West Coast) and Europe. Currently they (1) expect their cloud customers (i.e., SaaS providers) to express a preference for the location where they want their application services to be hosted and (2) do not provide seamless/automatic mechanisms for scaling their hosted services across multiple, geographically distributed data centers. This approach makes it difficult for cloud customers to determine in advance the best location for hosting their services as they may not know the origin of consumers of their services. In addition, no single cloud infrastructure provider will be capable to establish their data centers at all possible locations throughout the world. As a result, cloud application service (SaaS) providers may not be able to meet QoS expectations of their service consumers originating from multiple geographical locations [14].

Besides, commercial cloud providers make specific commitments to their customers by signing service level agreements (SLAs) [16]. SLA is a contract between the service provider and the customer to describe the provider's commitment and to specify penalties if those commitments are not met. For example, Amazon S3 (storage) cloud service SLA includes 99.9% uptime during a monthly billing cycle. The service credit percentage is either 10% or 25%, depending on the percentage of uptime. The EC2 (compute) SLA includes 99.95% availability during a service YEAR. The service credit in this case is 10%. The Microsoft Windows Azure SLA includes 99.95% uptime for Computing Connectivity and 99.9% uptime for Database availability, Storage availability, and Service availability. Financial penalties are made in the form of credits, which are based on the percentage of downtime. However, over the past 2 years, cloud service outages (i.e. service unavailability) have occurred frequently with many of the cloud providers, including Amazon, Google and Microsoft. Table 1.1 shows fail over records from some of the cloud provider system. These are the significant downtime incidents. Reliance on cloud can cause real problems when time is money.

Table 1.1 Outages in different cloud services

Cloud service and outage	Duration	Date	Implications
Google GMail	30 h	October 16, 2008	Users could not access their emails [91]
Google Gmail and Google Apps	24 h	August 15, 2008	Those affected by the outage received a 502 Server error when trying to log in to Gmail and Google Apps [91]
FlexiScale: core network failure	18 h	October 31, 2008	All services were unavailable to customers [86]
Amazon S3	6–8 h	July 20, 2008	Users could not access the storage due to single bit error leading to gossip protocol blowup [2]
Google Network	3 h	May 14, 2009	The vast majority of Google services became unavailable, including Gmail, YouTube, Google News, and even the google.com home page. The outage affected about 14% of Google users worldwide [73]
Google News	1.5 h	May 18, 2009	Users saw a 503 server error, along with a message to try their requests again later [70]
Google News	2 h	September 22, 2009	Many users experienced difficulties accessing Google News [26]
Google GMail	2 h	September 1, 2009	Users could not access their emails [29]
Amazon EC2	8 h	December 10, 2009	Customers experienced a loss of connectivity to their service instances [65]
Microsoft Sidekick	6 days	March 13, 2009	The massive outage left Sidekick customers without access to their calendar, address book, and other key aspects of their service [91]
Microsoft Azure	22 h	March 13, 2009	The outage occurred before the service came out of beta. The outage left people without access to their applications [91]
Netsuite	30 min	April 27, 2010	The company's cloud applications were inaccessible to customers worldwide [64]

The main reason for these cloud service outages is to face difficulty in handling cloud "bursting" (using remote resources to handle peaks in demand for an application). Thus, these repeated outages as depicted in Table 1.1 not only just lose the revenue from some services, but also imply a loss of reputation and therefore, a loss of future customers for cloud providers [48]. This has further implications to damage the credibility of the cloud. Enterprises that are considering the pros and cons of moving office productivity or communications to the cloud have reasoned to be concerned when the poster child of cloud Computing cannot provide reliable availability.

To support heterogeneous collaborative cloud services with different QoS targets, handle unpredictable and geographical changes in system workload, avoid SLA violation (i.e. cloud service outages) and reduce high operational costs, there is a need for a federation or collaboration platform among various cloud providers. Analysis of the current research efforts [14,36,51,61,63,69,72,74,81] in this context reveals that there has been only modest progress on the frameworks and policies required to achieve collaboration model among cloud providers. Furthermore, these researches mainly focus on supply chain collaboration (SCC) models, in which CPs leverage cloud services from other CPs for seamless provisioning. For example, Reservoir project [74] describes the model and architecture of federated cloud computing. It is using grid interfaces and protocols to realize interoperability between the clouds or infrastructure providers. IBM Altocumulus, a cloud middleware platform from IBM Almaden Services Research, is being developed to solve this very issue of managing applications across multiple clouds [63]. Aneka-Federation proposed in [72] provides a decentralized overlay for federation of Enterprise clouds. It considers only one kind of enterprise clouds in the federation.

The reasons for this lack of progress are due to the complexity of the technological problems (e.g. application service behavior prediction, flexible mapping of heterogeneous services to resources, and integration), legal, and commercial operational issues (e.g. economic models and scalable monitoring of system components) that need to be solved in the practical context.

Besides, in the near future, we can expect that hundreds of cloud providers will compete to offer services and thousands of users also compete to receive the services to run their complex heterogeneous applications on cloud computing environment. For example, emerging cloud applications like social networking, gaming portals, business applications, media content delivery, and scientific work flows need different composition, configuration, and deployment requirements. Quantifying the performance of scheduling and allocation policies in a real cloud environment for these applications under different conditions are extremely challenging because: (a) clouds exhibit varying demand, supply patterns, and system size; and (b) users have heterogeneous and competing QoS requirements [15]. In these open cloud collaboration scenarios; the existing federation/collaboration models are not applicable. In fact, while clouds are typically heterogeneous and dynamic, the existing federation models (i.e. SCC) are designed for static environments where a priori agreements among the parties are needed to establish the federation [18,34].

To move beyond these shortcomings, this book establishes the basis for developing advanced and efficient collaborative cloud service solution, in which cloud providers (smaller, medium, and large) of complementary service requirements will collaborate dynamically to gain economies of scale and enlargements of their capabilities to meet QoS targets of consumers. We term the technology for this interconnection and inter-operation of cloud providers in open cloud collaboration as "*Dynamic cloud Collaboration*" (DCC), as defined in the following:

Dynamic Cloud Collaboration (DCC) is formed by a set of autonomous smaller, medium and large cloud providers (CP_1, CP_2 ..., CP_n), which collaborate dynamically for sharing their own local resources (services) with others to fulfill the complementary service requirements of consumers with high QoS standards and minimum service cost. Each CP must agree with the resources/services contributed by other providers against a set of its own policies.

1.3 Significance of Dynamic Cloud Collaboration

DCC is a viable business platform where clouds can cooperate together accomplishing trust contexts and providing new business opportunities such as cost-effective assets' optimization, power saving, and on-demand resources provisioning. In fact, the existing federation models are designed for static environments where a-priori agreements among the parties are needed to establish the federation. In a DCC platform:

1. Each CP can share its own local resources/services with other CPs and so can dynamically expand or resize their provisioning capability based on sudden spikes in workload demands.
2. Each provider can maximize its profits by offering existing service capabilities to collaborative partners so that they may create a new value-added collaborative service by mashing-up existing services and these capabilities can be made available and tradable through a service catalog for easy exchange to consumers.
3. The capacity of delivering on demand, cost-effective, and QoS aware services of every CP increases without maintaining or administering any additional computing nodes, services or storage devices.
4. The reliability of a cloud service (i.e. handling of cloud service outage) is enhanced as a result of multiple redundant clouds that can efficiently tackle a disaster condition, ensuring business continuity.

Thus a DCC platform allows cloud providers to cooperatively achieve greater scale and reach, as well as service quality and performance, than they could otherwise attain individually. Its significance can be better understood by the following examples:

• Emerging cloud applications like Social networks (e.g. Facebook, MySpace etc.) deployed on a cloud provider, serve dynamic content to millions of users, whose access and interaction patterns are hard to predict. In addition, the dynamic creation of new plug-ins by independent developers may require heterogeneous resource requirements (high computing power, large storage, high-bandwidth capacity, etc.) which may not be provided by the hosting cloud provider since each cloud provider has specialization in their service provisioning (e.g. compute cloud, storage cloud, etc.). In these situations, load spikes (cloud bursting) can take place at different locations at any time, for instance, whenever new system features become popular or a new plug-in application is deployed. Thus result

in an SLA violation and end up incurring additional costs for the cloud provider [14]. This necessitates building mechanisms for dynamic collaboration of cloud providers for seamless provisioning of complementary service requirements.
- Other example applications that need DCC are massively multiplayer online role-playing games (MMORPGs). World of Warcraft [84], for example, currently has 11.5 million subscribers; each of whom designs an avatar and interacts with other subscribers in an online universe. Second Life is an even more interesting example of a social space that can be created through DCC. Any of the 15 million users can build virtual objects, own virtual land, buy and sell virtual goods, attend virtual concerts, bars, weddings, and churches, and communicate with any other member of the virtual world [88]. These MMORPGs certainly require different complementary cloud resources/services which cannot be provided by a single cloud provider. Thus, this necessitates building mechanisms for seamless collaboration of various cloud providers supporting dynamic scaling of applications across multiple domains in order to meet QoS targets of MMORPGs customers.

Such resource/service sharing and cooperation across different cloud providers may vary in terms of the purpose, scope, incoming application heterogeneity, and duration.

1.4   Research Challenges

This book tackles the research challenges in relation to the development of advanced, high performance, and cost effective dynamic collaborative cloud service solutions by enabling coordination and cooperation among multiple cloud provider services. We identify and investigate the following four key research issues:

When to collaborate: The circumstances under which a DCC arrangement should be performed. A suitable cloud market model is required that can enable and commercialize dynamic collaboration of cloud capabilities, hiring resources and assembling new services. Such a market can provide principles for efficient resource allocation depending on user QoS targets and workload demand patterns. It also offers opportunities to consumers as well as service-sharing incentives for cloud providers.

How to collaborate: The architecture that virtualizes multiple cloud providers. Such architecture must specify the interactions among entities and allow for divergent policies among participating providers.

Whom to collaborate with: The decision making mechanism used for choosing suitable partner cloud providers to collaborate with. A large number of conflicts may occur in a market-oriented dynamic collaboration platform when negotiating among cloud providers. One reason for the occurrence of the large number of conflicts is that each cloud provider must agree with the resources/services contributed by other providers against a set of its own policies in DCC [66, 67]. Another reason is the inclusion of high collaboration costs (e.g., network establishment, information transmission, capital flow) by the providers with their bidding prices as they do not know with whom they need to collaborate after winning an auction. Usually partner selection problem (PSP) needs a large quantity of factors (quantitative or qualitative ones) simultaneously and has been proven to be NP-hard [59] or NP-complete [71].

How to demonstrate dynamic collaborating applicability: The process to show the usefulness of explored strategies using simulation study.

1.5   Contributions of the Book

The main contributions of this book are as follows:

- Architectural framework and principles for the development of DCC. It describes the components, architectural features, use cases, and formation of dynamic collaborating arrangements. In addition, we present the utility of DCC to measure its content-serving ability as compared to the existing static cloud collaboration. It captures the heterogeneous application requirements in the system and helps to reveal the true propensities of participating cloud providers in collaboration. Through extensive simulations, interesting observations on how the utility is varied for different system parameters are presented. The challenges and core technical issues to implement DCC are also discussed, thus establishing the basis to develop necessary enabling techniques.
- A novel combinatorial auction (CA)-based cloud market model with a new auction policy called CACM is proposed to facilitate DCC platform among cloud providers. To address the issue of conflicts' reduction among cloud providers, the existing auction policy of CA [12, 13, 31] is modified. In existing CA model, each bidder (cloud providers) is allowed to compete for a set of services separately. After the bidding, the winning bidders need to collaborate with each other and so a large number of conflicts may occur when negotiating among providers. The new auction policy in the CACM model allows a cloud provider to collaborate dynamically with suitable partner cloud providers to form a group before joining the auction and to publish their group bid as a single bid to fulfill the service requirements completely, along with other providers, who publish separate bids to fulfill the service requirements partially. This approach can create more opportunities to win auctions since collaboration cost, negotiation time and conflicts among cloud providers can be reduced. We implement the proposed CACM model in a simulated environment and study its economic efficiency with the existing CA model.
- To find a good combination of cloud service partners for forming groups, a multi-objective (MO) optimization model for quantitatively evaluating the partners considering individual information (INI) and past relationship information (PRI) with collaboration cost (CC) optimization among partners is proposed. In the

existing approaches for partner selection, the INI is mostly used, while the PRI with collaboration cost optimization among partners is typically overlooked.

- To solve the MO optimization model for partner selection, a general framework of multi-objective genetic algorithm (MOGA) that uses INI and PRI of cloud providers, called MOGA-IC, is also presented. We develop MOGA-IC using the two popular MOGAs- non-dominated sorting genetic algorithm (NSGA-II) [32] and strength Pareto evolutionary genetic algorithm (SPEA2) [102], to find an appropriate diversity preservation mechanism for selecting operators to enhance the yield of Pareto-optimal solutions during optimization with multiple conflicting objectives. A numerical example is presented to illustrate the proposed MOGA-IC with NSGA-II and SPEA2 for cloud partner selection. In addition, we develop MOGA-I (MOGA using individual information), an existing partner selection algorithm, to validate the performance of MOGA-IC in the CACM model. Simulation experiments are conducted to show the effectiveness of the proposed MOGA-IC compared to that of MOGA-I in terms of satisfactory partner selection and conflicts' minimization.

1.6   Summary

To summarize, the work presented in this book is in line with the current trends in cloud providers to shift towards cloud collaboration or federation model [39], which allows consumers to have collaborative cloud services without having to build a dedicated infrastructure.

Chapter 2
Related Work

Abstract In this chapter, first, we provide an overview of present cloud collaboration or federation initiatives. Second, we discuss existing works on market-oriented cloud models. Finally, existing collaborator or partner selection approaches in different areas are presented.

2.1 Cloud Federation or Collaboration Initiatives

Cloud collaboration or federation is gaining popularity in the research community, due to its flexibility and effectiveness to improve performance for end-users and to achieve pervasive geographical coverage and increased capacity for a cloud provider. In this section, we provide a comparative analysis of research related to cloud collaboration or federation in order to ascertain the feasibility of our architecture and position it as the basis to address the shortcomings of prior initiatives.

The idea of federating systems was already present in the Grid. For instance, works such as [10] and [80] use federation in order to get more resources in a distributed Grid environment. The application of federation or collaboration in the cloud was initially proposed within the Reservoir project. In particular, B. Rochwerger et al. in [74] proposed the Reservoir model and architecture for open federated cloud computing. Reservoir is a European Union FP7 funded project aim to facilitate an open, service-based, on-line economy, where resources and services are transparently provisioned and managed across clouds on an on-demand basis at competitive costs with high quality of service. The authors attempt to use grid interfaces and protocols to realize interoperability between the clouds or infrastructure providers, but their work is in the model stage.

Ranjan et al. describe a decentralized overlay for federation of enterprise clouds called Aneka-Federation [72], Aneka is a .NET-based service-oriented resource management platform, which is based on the creation of containers that host the services, and it is in charge of initializing services and act as a single point for

M.M. Hassan and E.-N. Huh, *Dynamic Cloud Collaboration Platform:*   13
A Market-Oriented Approach, SpringerBriefs in Computer Science,
DOI 10.1007/978-1-4614-5146-4_2, © The Author(s) 2013

interaction with the rest of the Aneka cloud. Moreover, it provides SLA support such that the user can specify QoS requirements such as deadline and budget. However, there is only one kind of enterprise cloud in the federation, portability and interoperability between different cloud providers is not mentioned in this work. Furthermore, no economic model is presented to commercialize the federation.

Buyya et al. [14] presents vision and challenges of Inter-cloud for utility-oriented federation of cloud computing environments. The authors proposed an architecture for Inter-Cloud environment, a cloud coordinator for exporting cloud services, a cloud broker for mediating between service consumers and cloud coordinators, a cloud exchange for match making service and finally, a software platform to implement cloud coordinator, broker, and exchange.

IBM Altocumulus [63], a cloud middleware platform from IBM Almaden Services Research, aims to solve the issue of managing applications across multiple clouds. It provides a uniform, service oriented interface to deploy and manage applications in various clouds and also provides facilities to migrate to instances across clouds using repeatable best-practice patterns.

Several other research efforts follow cloud federation initiative. Erik Elmroth et al. [36] proposed technology neutral interfaces and architectural additions for handling placement, migration, and monitoring of virtual machines (VMs) in federated cloud environments. The interfaces presented to adhere to the general requirements of scalability, efficiency, and security in addition to specific require-ments related to the particular issues of interoperability and business relationships between competing cloud computing infrastructure providers.

Several standardization projects are going on for developing interoperable cloud interfaces, such as the OCCI [68] working group at the Open Grid Forum. Also the open cloud manifesto (OCM) initiative [http://www.opencloudmanifesto.org/] launched by IBM is trying to provide a core set of principles for interoperability between cloud providers. Cisco, Sun, and SAP all signed it and support it. The cloud computing Interoperability Forum (CCIF) [http://www.cloudforum.org/] is also formed in order to enable a global cloud computing ecosystem. A key focus will be placed on the creation of a common agreed upon framework/ontology that enables the ability of two or more cloud platforms to exchange information in an unified manor. In addition, another standardization effort called open cloud consortium (OCC) [http://opencloudconsortium.org/] supports the development of standards for cloud computing and frameworks for interoperating between clouds, develops benchmarks for cloud computing, and supports reference implementations for cloud computing.

Analysis of the existing works mentioned above reveals that these works mostly focus on Supply Chain Federation or Collaboration Model in which cloud providers will leverage cloud services from other cloud providers for seamless provisioning [18]. However, in the near future, we can expect that hundreds of cloud providers will compete to offer services and thousands of users also compete to receive the services to run their complex heterogeneous applications on cloud computing environment. In these open cloud collaboration scenarios, the existing supply chain federation models are not applicable. In fact, while clouds are typically heterogeneous and dynamic, the existing federation models are designed for static environments where a-priori agreements among the parties are needed to establish the federation [18, 34]. Thus there is a need for an advanced and efficient collaborative cloud service solutions in which cloud providers (smaller, medium, and large) of complementary service requirements will collaborate dynamically to gain economies of scale and enlargements of their capabilities to meet QoS targets of heterogeneous cloud service requirements. But there has been only modest progress on the frameworks and policies required to achieve this collaboration model among cloud providers.

In [34], an architecture for Cross-cloud system management is proposed to facilitate the management of compute resources from different cloud providers in a homogeneous manner. The primary goal is to provide the flexibility and adaptability promised by the cloud computing paradigm, whilst enabling the benefits of cross-cloud resource utilization to be realized. Our approach is significantly different as they do not capture the heterogeneity of the consumer service requirements and only consider homogeneous service requirements.

In [18], the authors proposed a solution based on the Cross-Cloud Federation Manager, a new component placeable inside the cloud architectures, allowing a cloud to establish the federation with other clouds according to a three-phase model: discovery, match-making and authentication. While this work is appealing, the circumstances under which the collaboration will be performed, the decision making mechanism to choose suitable partners and the strategy to motivate, and form collaboration are unexplored in this work.

2.2   Market Models for Cloud Collaboration

Very few approaches [6, 14, 16, 17, 48, 49] are proposed in the literature regarding the cloud market model. In [48], authors focus on a complete characterization of provider's federation in the cloud, including decision equations to outsource resources to other providers, rent free resources to other providers (i.e. insourcing), or shutdown unused nodes to save power. A first approach introducing this idea is presented in [17] where they state some factors such as provider occupation and maintaining costs in order to dimension a cloud provider and when to outsource to a federated provider. But these approaches just focus on to decide when to move tasks to a federated provider based on economic criteria.

In [16], the authors present a vision of twenty-first century computing, describe some representative platforms for cloud computing covering the state-of-the-art and provide the architecture for creating market-oriented clouds for resource management. In another paper [14], the same authors use this model to create a market for supply chain cloud federation model. The authors proposed an auction based general cloud market for trading cloud services and resource management since auctions can provide a solution for scalable resource economies. However, this market model cannot be directly applicable in creating a DC platform among

CPs since the DC platform deals with a combinatorial allocation problem. So in next section, we describe the combinatorial auction model for our DC platform.

2.2.1   Combinatorial Auction

An auction is a mechanism to allocate a set of goods to a set of bidders based on the bids and asks. The word auction usually refers to a single sided mechanism (single buyer-multiple sellers or single seller-multiple buyers). There are three types of auctions- one-sided auction, double-sided auction and combinatorial auction (CA) [12, 13, 31]. Compared to other approaches, CA is efficient and can maximize revenue, and it is the appropriate market mechanism for our cloud market to enable dynamic collaboration.

A combinatorial auction is an auction where the bidders are allowed to submit bids on combinations or subsets of items to an auctioneer. Combinatorial auctions are one-sided combinatorial mechanisms, which could be either forward (single seller and multiple buyers) or reverse (single buyer and multiple sellers). The auctioneer decides (after one or more rounds or after a certain amount of time depending upon the design) to accept some of the bids and to allocate the items accordingly to the bidders.

2.2.1.1   Generalized Vickrey Auction

GVA is a sealed-bid auction where bids are sealed and hidden from other bidders. However, in this auction form, winner of the auction pays only the amount of second highest bid. It is a very general method for designing truthful mechanisms devised by Clarke and Groves [27]. When applied to combinatorial auctions, it generalizes the second price sealed bid auction of Vickrey [89] and is therefore called the generalized Vickrey auction (GVA). The GVA mechanism satisfies the following desirable properties:

- *Allocative Efficiency*: The allocation determined by GVA is efficient, that is, it maximizes total valuation across agents, given rational agent strategies.
- *Strategy Proofness*: Truth revelation is optimal for an agent irrespective of the strategies followed by other agents.
- *Individual Rationality*: The expected utility from participation is non-negative with a rational strategy.
- *Weak Budget Balance*: Each agent makes a non-negative payment to the auction-eer, so the net revenue collected by auctioneer is non-negative.

GVA is an important building block for combinatorial auctions. Several CA mechanisms are based on GVA. Examples of such mechanisms are discussed in [79].

2.2.1.2 The Winner Determination Problem

In a combinatorial auction in its most general form, bidders can bid whatever amount they please on any subset of items in which they are interested. The problem of deciding which bidders should get what items in order to maximize the total winning bid value is called the winner determination problem. Furthermore, we assume that at most one bid per bidder can be accepted. Indeed, a bidder's bid for the whole set of items he wins might well be smaller than the sum of bids for the underlying subsets.

Suppose for instance a bidder j expresses the following bids:

$b_j(\{1\}) = 3$, $b_j(\{2\}) = 2$, and $b_j(\{1,2\}) = 4$

Accepting both the bid on item 1 and the bid on item 2 leads to a combined bid of $2+3 = 5$, whereas this bidder intended to bid no more than 4 for the combination of items 1 and 2.

The following formulation is most commonly used to represent this problem (for a single-unit setting). We use B to represent the set of bidders indexed by j and G for the set of items indexed by i. We use $b_j(S)$ to denote the bid by bidder j on the set of items $S \in \Omega_j \subseteq 2^G$, where Ω_j is the set of sets in which bidder j is interested. The binary variable $y(S,j)$ indicates whether bidder j wins the set $S(y(S,j) = 1)$ or not (otherwise).

maximize

$$\sum_{j \in B} \sum_{S \in \Omega_j} b_j(S) y(S,j) \tag{2.1}$$

subject to

$$\sum_{S \in \Omega_j : S \supseteq \{i\}} \sum_{j \in B} y(S,j) \leq 1 \qquad \forall i \in G \tag{2.2}$$

$$\sum_{S \in \Omega_j} y(S,j) \leq 1 \qquad \forall j \in B \tag{2.3}$$

$$y(S,j) \in \{0,1\} \qquad \forall S \in \Omega_j, \forall j \in B \tag{2.4}$$

The first set of constraints (2.2) enforces that no item can be auctioned more than once. The second set of constraints (2.3) ensures that there is at most one winning bid per bidder. The winner determination problem is shown to be NP-hard, even if every bidder bids only on subsets of size and all bids have a value equal to 1 [87]. So we utilize secured generalized Vickrey auction (SGVA) [85] to address the CACM model problem and use dynamic graph programming [98] for winner determination algorithm.

2.2.2 Limitations of Existing CA Model

In CA-based market model, the user/consumer can bid a price value for a combination of services, instead of bidding for each task or service separately and each bidder or service provider is allowed to compete for a set of services. This auction policy of the CA-based market model is not fully suitable to meet the requirements of a DC platform. If the existing auction policy of CA model is applied, each bidder (CP) is allowed to compete for a set of services separately. After the bidding, the winning bidders need to collaborate with each other. As we mentioned earlier, a large number of conflicts may occur when negotiating among providers in the DC platform [66,67]. So the CA-based market model cannot address the issue of conflict minimization among the CPs in a dynamic collaboration platform.

We propose to modify the existing auction policy of CA that allows the CPs to publish their bids collaboratively as a single bid in the auction by dynamically collaborated with suitable partners. This approach can help to minimize the conflicts and collaboration cost among CPs as they know each other very well in the group and also create more chances to win the auction. However, finding appropriate partners to make a group is a complex NP hard problem [59]. We need to find a good partner selection algorithm that can minimize conflicts as well as collaboration cost among providers.

The proposed approach for group formation to provide combined services has some similarity with "Bundle Search" problem in the electronic market where a buyer needs to buy multiple goods as a bundle from different sellers/suppliers [21]. The partnerships between suppliers result in different bundles having different discounts. Typical applications are travel packaging, software, PC peripherals, etc. However, there are differences between partner selection as well as negotiation mechanisms used in bundle search problem and dynamic cloud collaboration scenario. In bundle search problem scenario, negotiation is done between two participants at a time. The agreement is signed between two participants, and they have distinct roles to play in reaching an agreement. The final agreement is signed between the consumer and the Integrator/Broker. In contrast, in case of dynamic collaboration, all participants have to contribute their resources as well as agree with the resources contributed by others and so have to sign the same agreement. So the negotiation mechanism used in bundle search problem cannot be used in the context of dynamic collaborations because of inherent multi-party nature of the negotiation. A distributed many-to-many negotiation protocol is used in case of dynamic cloud collaboration.

2.3 Collaborator/Partner Selection Algorithms

Collaborator or Partner selection problem (PSP) is a complex problem, which usually needs a large quantity of factors (quantitative or qualitative ones) si-multaneously, and has been proven to be NP-hard [59] or NP-complete [71]. Also partner selection problem for CPs in dynamic collaboration environment is different from other partner selection problems in areas like virtual enterprise [25, 44, 52, 90, 92, 94], dynamic alliances [37, 42], international joint ventures [50], supply chain [20, 22, 38, 41, 75, 95] or production networks [43, 53]. In dynamic cloud collaboration case, each CP partner must share its own resources/services with

Table 2.1 Partner selection criteria in different areas

Area	Relevant studies	Partner selection criteria
Virtual enterprise	Huang et al. [52], Ye and Li [94], Jarino and Salo [56], Zhong et al. [101], Wang et al. [90]	Price, transportation cost, quality, risk, time and due date
Supply chain	Yeh and Chuang [95], Farahani et al. [41], Che et al. [22]	Production cost, transportation time, quality, yield and air pollution treatment cost
Alliance formation	Feng et al. [42], Fan and Feng [39], Famuyiwa et al. [38], Emden et al. [37]	Cost, quality, time, capability, knowledge, resource complementarity, overlapping knowledge bases, cultures, goal correspondence, leadership, academic influence, and interpersonal communication
Production network	Cheng et al. [24], Fischer et al. [43]	Price, processing cost, penalty cost and production load
International joint venture	Chen et al. [23], Hajidimitriou and Georgiou [50]	Product price, transportation cost, profit goal, financial ratio index and collaborative filtering

others and provide some collaborative service and also agree with resources/services contributed by other providers against a set of its own policies. So a large number of conflicts may happen among CPs and thus it is very difficult to choose appropriate CP partners in dynamic cloud collaboration environment.

There has been much research regarding the attributes (or criteria) for partner selection. Some representative partner selection criteria in different areas are shown in Table 2.1.

From Table 2.1, we can see that in the existing studies on partner selection, the individual information (INI) is mostly used, but the past relationship information (PRI) with collaboration cost optimization between partners is overlooked. In fact, the PRI [30, 40] with collaboration cost optimization is very important in partner selection for CPs in dynamic collaboration environment. The success of past relation with fewer collaboration costs between participating CPs may reduce uncertainty and conflicts, short adaptation duration, and also help to the performance promotion. So it is difficult to adopt directly the existing methods to solve the partner selection problem of CPs using the INI and PRI with collaboration cost optimization into consideration.

There has been also much research addressed the algorithms for partner selection like fuzzy decision-making algorithms [42, 43], quantitative algorithms [56, 92], genetic algorithms [24, 41, 44, 55, 90, 95] algorithms and Hybrid algorithms [25, 53].

The individual information considered in the existing methods is associates with a single candidate partner. However, the PRI with collaboration cost optimization additionally considered in this paper, is shared by pairwise partners. Additionally, individual and past relationship with collaboration cost optimization utilities should be integrated to obtain the overall ranking value of each candidate partner. Therefore, existing methods could not be directly used to solve the problem addressed in this paper. A novel method needs to be investigated for cloud partner selection in dynamic cloud collaboration platform.

2.3.1 Multi-objective Optimization and Genetic Algorithms

In the CP partner selection problem (PSP), multi-objective (MO) optimization is preferable because it provides a decision-maker (pCP) with several trade-off solutions to choose from. Also the MO formulations are practically required for concurrent optimization that yields optimal solutions that balance the conflicting relationships among the objectives. MO optimization yields a set of Pareto optimal solutions, which is a set of solutions that are mutually non-dominated [32, 102]. The concept of non-dominated solutions is required when comparing solutions in a multi-dimensional feasible design space formed by multiple objectives. The multi-objective partner selection optimization problem is expressed as a vector of functions as follows:

Maximize/Minimize $z = f_1(x), f_2(x), \ldots\ldots, f_w(x)$ where Z, w, $f_w(x)$ and x are the multi-objective vector function, the number of objective functions, the wth objective function, and a set of design variables, respectively.

In terms of minimization of all objectives, a feasible solution x_1 is said to dominate another feasible solution x_2 ($x_1 \succ x_2$) if an only if $f_w(x_1) \leq f_w(x_2)$ for $l = 1, \ldots, L$ and $f_w(x_1) < f_w(x_2)$ for at least one objective function w [103]. A solution is said to be *Pareto-optimal* if it is not dominated by any other solution in the solution space. The set of all such feasible non-dominated solutions in a solution space is termed the Pareto optimal solution set. For a given Pareto-optimal solution set, the curve made in the objective space is called the *Pareto front* (see Fig. 2.1). When two conflicting objectives are present there will always be a certain amount of sacrifice in one objective to achieve a certain amount of gain in the other when moving from one Pareto solution to another. So often it is preferred to use a Pareto optimal solution set rather than being provided with a single solution, because the set helps effectively understand the trade-off relationships among conflicting objectives and make informed selections of the optimal solutions.

The use of multi-objective genetic algorithms (MOGAs) provides a decision-maker with the practical means to handle MO optimization problems. Single-objective Genetic Algorithms (GAs) that can be modified to solve MO optimization problems and find Pareto optimal sets in a single run are usually called multi-objective GAs (MOGAs) [60]. Most of the MOGAs do not require artificial adjustments such as priority, scaling, or weighting coefficients for the objective functions [60]. An additional advantage is that the crossover and mutation operators may be modified to exploit the structural features of preferable solutions.

Over the years, a number of MOGAs have been developed [28, 32, 77, 96, 102]. For cloud partner selection problem, one important issue is to find an appropriate diversity preservation mechanism in selection operators by incorporating density

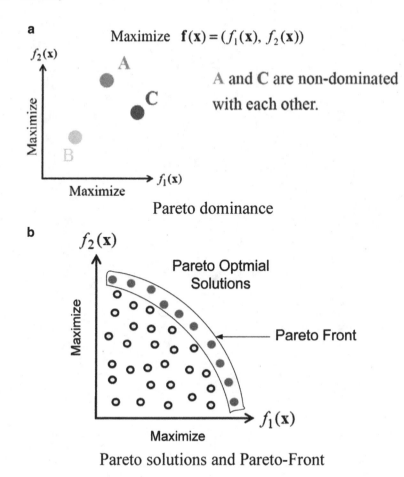

Fig. 2.1 Pareto optimality concept. (**a**) Pareto dominance, (**b**) Pareto solutions and Pareto-Front

information to enhance the yield of Pareto optimal solutions during optimization with multiple conflicting objectives. In this case, two popular MOGAs are used, namely the fast non-dominated sorting GA (NSGA-II) [32] and the strength Pareto evolutionary algorithm (SPEA2) [102], since both two algorithms includes effective mechanisms for preserving diversity and can yield better Pareto optimal solution sets.

2.4 Summary

This chapter first presents the current research efforts of cloud collaboration or federation initiatives. Second, it discusses research works on market-oriented cloud models and identifies the limitations of current market approaches for DC platform. Third, it presents existing collaborator or partner selection approaches in different areas with various partner selection criteria.

Chapter 3
Architectural Framework and Market Model for Dynamic Cloud Collaboration

Abstract Existing cloud providers, operating in isolation, are often prone to Service Level Agreement violations and resources over-provisioning in order to ensure high-quality services to end-users, thus incurring extensive operational cost and labor. As mentioned in Chap. 1, dynamic cloud collaboration (DCC) is an approach to reduce expenses and avoid adverse business impact. It is formed by a set of autonomous cloud providers who cooperate through a mechanism to share resources while enjoying larger scale and reach. This chapter first presents the architecture that establishes the basis to form DCC. Finally, it describes the proposed combinatorial auction (CA)-based cloud market model called CACM that enables and commercializes a DCC platform.

3.1 Introduction

A dynamic cloud collaboration (DCC) platform exhibits a negotiated resource sharing relationship among different cloud providers. It can help CPs to maximize their profits by offering existing services' capabilities to cooperative business partners. These capabilities can be available and tradable through a service catalog for easy mash-up to provide new value-add collaborative cloud services to consumers. Also the DC platform can enable a CP to handle cloud bursting by redirecting some load to collaborators. Figure 3.1 shows a virtual organization (VO) based DCC platform consisting of resources/services spanning multiple cloud providers.

The formation of a DCC can be initiated by a cloud provider, which recognizes a good business opportunity in forming a dynamic collaboration platform with other CPs in order to provide a set of complementary services to consumers. The initiator is called the *primary CP* (pCP), while other CPs who share their resources/services in a DCC platform are called *collaborating or partner CPs*. Users interact transparently with the DCC platform by requesting services through a service catalog of the pCP. The CPs offer capabilities/services to consumers with a full consumption specification formalized as a standard SLA. The requested service

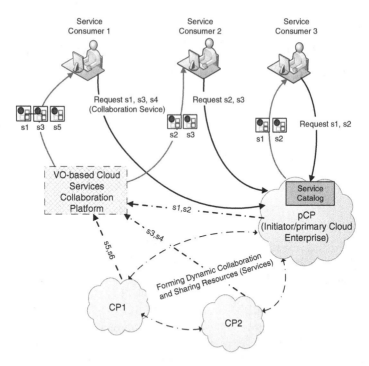

Fig. 3.1 A formed VO based cloud services collaboration platform

requirements (single, multiple or collaborative cloud services) are served either directly by the pCP or by any collaborating CP within the DCC. Suppose that a pCP can provide two services s1 and s2 and CP1 and CP2 can provide services s3, s4 and s5, s6 respectively, as shown in Fig. 3.1. The request for collaborative services s1, s3, s5 or s2, s3 can be served by the DCC platform. In case of services s1 and s2, the pCP can directly deliver the services. Thus, a DCC platform can deliver on demand, reliable, cost-effective, and QoS aware services based on virtualization technologies while ensuring high QoS standards and minimizing service costs.

3.2   System Architecture of Dynamic Cloud Collaboration

Figure 3.2 presents the cooperative architecture of a system for DCC. The terminologies used to describe the system are listed in Table 3.1.

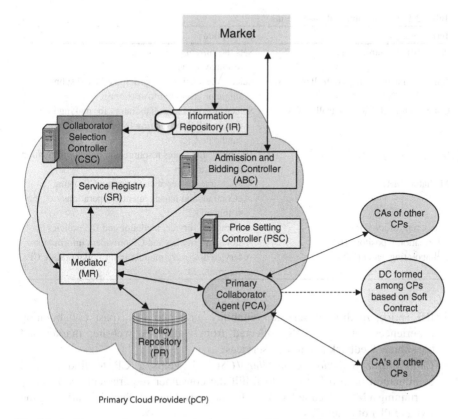

Fig. 3.2   Architecture of a system to assist the creation of dynamic cloud collaboration

3.2.1   Architectural Components

In the following we describe the core components of the proposed architecture along with their responsibilities:

- *Price Setting Controller (PSC)*: A CP is equipped with a PSC which sets the current price for the resource/service based on market conditions, user demand, and current level of utilization of the resource. Pricing can be either fixed or variable depending on the market conditions.
- *Admission and Bidding Controller (ABC)*: It selects the auctions to participate in and submits single or group bid based on an initial estimate of the utility. It needs market information from the information repository (IR) to make decisions which auction to join.
- *Information Repository (IR)*: The IR stores the information about the current market condition, different auction results and consumer demand. It also stores

Table 3.1 List of commonly used terms

Terminology	Description
Price setting controller (PSC)	Sets the current price for the resource/service
Admission and bidding controller (ABC)	Selects the auctions to participate in and submits single or group bid to auctioneer
Collaborator selection controller (CSC)	Selects appropriate collaborators for making groups using a MOGA utilizing INI and PRI of other candidate CPs
Service registry (SR)	Discovers and stores resource and policy information to local domain
Mediator (MR)	Responsible for negotiation among CPs using eContract and management of operations within a DC
Policy repository (PR)	A storage of service, mediator and DC policies
Information repository (IR)	A storage of market and CP providers information
Collaborating agent (CA)	A service discovery module in the collaborating CPs environment

INI (price, quality of service, reliability etc.) and PRI (past collaboration experiences) of other CPs collected from each CPs website, market and consumers feedback about their services.

- *Collaborator Selection Controller (CSC)*: It helps a CP to find a good combination of collaborators to fulfill the consumer requirements completely by running a MOGA called MOGA-IC (described later in Sect. 5.3) utilizing the INI and PRI of other CPs.
- *Mediator (MR)*: The MR controls which resources/services to be used for collaborative cloud services of the collaborating CPs, how this decision is taken, and which policies are being used. When performing DC, the MR will also direct any decision making during negotiations, policy management, and scheduling. A MR holds the initial policies for DC formation and creates an eContract and negotiates with other CPs through its local Collaborating Agent (CA).
- *Service Registry (SR)*: The SR encapsulates the resource and service information for each CP. In the case of DC, the service registry is accessed by the MR to get necessary local resource/service information. When a DC is created, an instance of the service registry is created that encapsulates all local and delegated external CP partners' resources/services.
- *Policy Repository (PR)*: The PR virtualizes all of the policies within the DC. It includes the MR policies and DC creation policies along with any policies for resources/services delegated to the DC as a result of a collaborating arrangement. These policies form a set of rules to administer, manage, and control access to DC resources and also helps to mash-up cloud services. They provide a way to manage the components in the face of complex technologies.
- *Collaborating Agent (CA)*: The CA is a policy-driven resource discovery module for DC creation and is used as a conduit by the MR to exchange eContract

with other CPs. It is used by a primary CP to discover the collaborating CPs (external) resources/services, as well as to let them know about the local policies and service requirements prior to commencement of the actual negotiation by the MR.

3.2.2 Dynamic Cloud Collaboration Life Cycle

DCC can be short-term wherein CPs operate to handle sudden load spikes, or long-term in which they explore the delivery of specialized services. A dynamic collaborating life cycle goes through six major steps, as described in the following:

Step 1: pCP finds a business opportunity in the market from IR and wants to submit collaborative bids as a single bid in the auction to address consumer requirements as it cannot provide all the service requirements

Step 2: The CSC is activated by the pCP to find a set of Pareto-optimal solutions for partner selection and it chooses any combination from the set to form groups and send this information to the MR.

Step 3: The MR obtains the resource/service and access information from the SR, whilst SLAs and other policies from the PR. It generates a eContract that encapsulates its service requirements on the pCP's behalf based on the current circumstance, its own contribution policies, prices of services (generated by PSC) and SLA requirements of its customer(s) and passes this eContract to the local Collaborating Agent (CA).

Step 4: The local CA of pCP carries out negotiations using a distributed many-to-many negotiation protocol proposed in [66] with the CAs of other identified partner CPs. An eContract is used in the negotiation process. It is used to capture the contributions as well as agreements among all participants. As the group members know each other very well, the number of conflicts will be less. So when all CPs (including the pCP) agree with each other, they make a soft contract among them. A soft contract guarantees that resources/services will be available if the group wins the auction.

Step 5: When pCP acquires all services/resources from its collaborator to meet SLA with the consumer, a DC platform is formed. If no CP is interested in such arrangements, DC creation is resumed from Step 2 with another Pareto-optimal solution.

Step 6: After the DC platform creation, the MR of pCP submits collaborative bids as a single bid to the market using the admission and bidding controller (ABC). If this group wins the auction, a hard contract is performed between each group member to firm up the agreement in DC. A hard contract ensures that the collaborating CPs must provide the resources/services according to the SLAs with consumers.

If some CPs win the auction separately for each service (few chances are available), the steps 3–5 are follows to form a DC platform among providers. But they make the hard contract in step 4 and in this case, a large number of conflicts may happen to form the DC platform.

An existing DC may need to either disband or re-arrange itself if any of the following conditions hold: (a) the circumstances under which the DC was formed no longer hold; (b) collaborating is no longer beneficial for the participating CPs; (c) an existing DC needs to be expanded further in order to deal with additional load; or (d) participating CPs are not meeting their agreed upon contributions.

3.3   Challenges to Realize Dynamic Cloud Collaboration

There are a number of challenges, both technical and non-technical (i.e. commercial and legal), that can block the rapid growth of DCC. In this section, we outline some of the common stoppers for the uptake of DCC.

- *Application service behavior prediction*: It is critical that the system is able to predict the demands and behaviors of the hosted services, so that it intelligently undertake decisions related to dynamic scaling or descaling of services over federated cloud infrastructures. Concrete prediction or forecasting models must be built before the behavior of a service, in terms of computing, storage, memory and network bandwidth requirements.
- *Flexible mapping of services to resources*: The process of mapping services to resources is a complex undertaking, as it requires the system to compute the best software and hardware configuration (system size and mix of resources) to ensure that QoS targets of services are achieved, while maximizing system efficiency and utilization. Consequently, there is an immediate need to devise performance modeling and market-based service mapping techniques that ensure efficient system utilization without having an unacceptable impact on QoS targets.
- *Economic models driven optimization techniques*: The market-driven decision making problem [6] is a combinatorial optimization problem that searches the optimal combinations of services and their deployment plans. Un-like many existing multi-objective optimization solutions, the optimization models that ultimately aim to optimize both resource-centric (utilization, availability, reliability, incentive) and user-centric (response time, budget spent, fairness) QoS targets need to be developed.
- *Integration and interoperability*: There is a need to look into issues related to integration and interoperability between the software on premises and the services in the cloud. In particular: (a) Identity management: authentication and authorization of service users; provisioning user access; federated security model; (b) Data Management: not all data will be stored in a relational database in the cloud, (c) Business process orchestration: how does integration at a business

process level happen across the software on premises and service in the cloud boundary? Where do we store business rules that govern the business process orchestration?

- *Scalable monitoring of system components*: Although the components that contribute to a federated system may be distributed, existing techniques usually employ centralized approaches to overall system monitoring and management. These centralized approaches are not an appropriate solution for this purpose, due to concerns of scalability, performance, and reliability arising from the management of multiple service queues and the expected large volume of service requests. Therefore, we advocate architecting service monitoring and management services based on decentralized messaging and indexing models.
- *Pricing of services*: Sustained resource sharing between participants in a collaborating arrangement must ensure sufficient incentives exist for all parties. It requires the deployment of appropriate pricing, billing, and management mechanisms. The key questions are:

  - What mechanisms are to be used for value expression (expression of content and service requirements and their valuation), value translation (translating requirements to content and service distribution) and value enforcement (mechanisms to select and distribute different content and services)?
  - How do cloud providers achieve maximum profit in a competitive environment, yet maintaining the equilibrium of supply and demand?

To meet aforementioned requirements of DCC, future efforts should focus on design, development, and implementation of software systems and policies for collaboration of clouds across network and administrative boundaries. In particular, the resource provisioning within these collaborated clouds will be driven by market-oriented principles for efficient resource allocation depending on user QoS targets and workload demand patterns. In the following section, we present our proposed combinatorial auction based cloud market (CACM) model to facilitate a DCC platform.

3.4 Combinatorial Auction Based Cloud Market Model

3.4.1 Market Architecture

The proposed CACM model to enable a DC platform among CPs is shown in Fig. 3.3. The existing auction policy of CA is modified in the CACM model to address the issue of conflicts minimization among CPs in a DC platform. The existing and new auction policy for CA model is shown in Fig. 3.4. The CACM model allows any CP to dynamically collaborate with appropriate partner CPs to form groups and to publish their group bids as a single bid to completely fulfill the

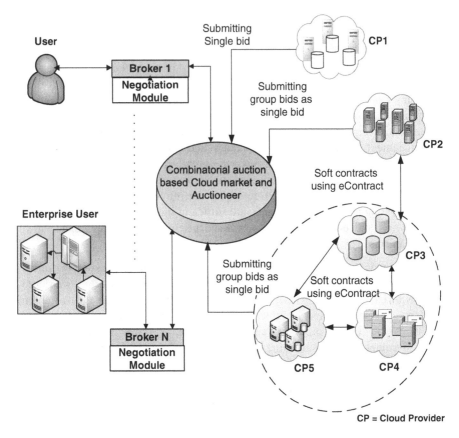

Fig. 3.3 Proposed CACM model to enable a DC platform among CPs

consumer service requirements while also supporting the other CPs to submit bids separately for a partial set of services. We use the auction scheme based on [33, 34] to address the CACM model. The main participants in the CACM model are brokers, users/consumers, CPs and auctioneers as shown in Fig. 3.3.

Brokers in the CACM model mediate between consumers and CPs. A broker can accept requests for a set of services or composite services requirements from different users. A broker is equipped with a negotiation module that is informed by the current conditions of the resources/services and the current demand to make its decisions. Consumers, brokers and CPs are bound to their requirements and related compensations through SLAs. Brokers gain their utility through the difference between the price paid by the consumers for gaining resource shares and that paid to the CPs for leasing their resources.

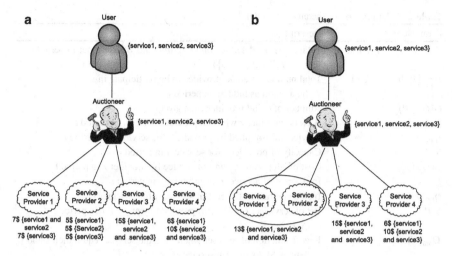

Fig. 3.4 Auction policies. (**a**) Existing auction policy of CA   (**b**) New auction policy in CACM

The users/consumers can be enterprise user or personal user. Consumers have their own utility functions that cover factors such as deadlines, fidelity of results, and turnaround time of applications. They are also constrained by the amount of resources that they can request at any time, usually by a limited budget. The users can bid a single price value for different composite/collaborative cloud services provided by CPs.

The CPs provide cloud services/resources like computational power, data storage, Software-as-a Service (SaaS), computer networks or infrastructure-as-a Service (IaaS). A CP participates in an auction based on its interest and profit. It can publish bid separately or collaboratively with other partner CPs by forming groups to fulfill the consumers' service requirements.

The responsibility of an auctioneer includes setting the rules of the auction and conducting the combinatorial auction. The auctioneer first collects bids (single or group bids) from different CPs participating in the auction and then decides the best combination of CPs who can meet user requirements for a set of services using a winner determination algorithm. We utilize secured generalized Vickrey auction (SGVA) [85] to address the CACM model problem and use dynamic graph programming [98] for winner determination algorithm.

In SGVA, the actual evaluation values of bidders are hidden by using homomorphic encryption. The evaluation value of a bidder is represented by a vector of cipher texts of homomorphic encryption which enables the auctioneer to find the maximum value and add a constant securely, while the actual evaluation values are kept secretly.

Table 3.2 Parameters for the auction model

Parameters	Description
$R = \{R_{jt}\,\vert\,j = 1\ldots n\}$	Total aggregated service unit requirements of consumer in period t $(t \in \{T, T-1, \ldots .0\})$
$P = \{P_{rt}\,\vert\,r = 1\ldots m\}$	Total number of cloud providers who participate in the auction as bidders in period t
$G(G \subseteq P)$	Number of cloud providers in a group
P_{rjt}	A cloud provider r who can provide service j in period t
s_{rjt}	Service unit supplied by provider r for service j in period t
$\tilde{S}_{rjt}$	Capacity of provider r for service j in period t
$C(s_{rjt})$	Cost of supplying s_{rj} unit of service by provider r in period t
$I_{C_{rjt}}$	Initial cost of service j for provider r in period t
η_{rj}	Increasing rate of the cost of service j for provider r
$S(P_{rt})$	Set of services $(j = 1, \ldots n)$ provided by any CP r in period t $(S(P_{rt}) \subset R)$
$\Omega_{\max}(R, Q)$	Payoff function of the user where R is the service requirements and Q defines SLAs of each service

3.4.2   System Model for Auction in the CACM

3.4.2.1   Single and Group Bidding Functions of CPs

For the convenience of analysis, the parameters for the auction models are shown in Table 3.2. Let M be the service cost matrix of any CP P_r and G be a group of CPs in P (i.e. $G \subseteq P$). To simplify the auction model, we assume that each CP can provide at most two services. The reason is that it is not possible for a CP to provide almost all kinds of services. The matrix M includes costs of a CP's own services as well as the collaboration costs (CC) between services of its own and other CPs. Figure 3.5 illustrates the matrix M.

We assume that P_r provides two services—CPU and Memory. Let $a_{jj}(j = 1\ldots n)$ be the cost of independently providing any service in M, $a_{ij}(i, j = 1\ldots n, i \neq j)$ be the CC between service i and j $(i, j \in S(P_r))$ and $a_{jk}(j, k = 1\ldots n, j \neq k)$ be the CC between services j and k $(j \in S(P_r)$ and $k \notin S(P_r))$. We set nonreciprocal CC between $S(P_r)$ services in M which is practically reasonable.

If CP P_r knows other providers or have some past collaboration experience with others, it can store true CC of services with other providers. Otherwise it can set a high CC for other providers. The CC of services with other providers in matrix M is updated when the providers finish a negotiation and collaboratively provide the services of consumers in the DC platform.

Now the "Bidding Function" or individual price of any CP say P_r who submits a bid separately to partially fulfill the customer service requirements at time t can be determined as follows:

Fig. 3.5 Cost matrix M

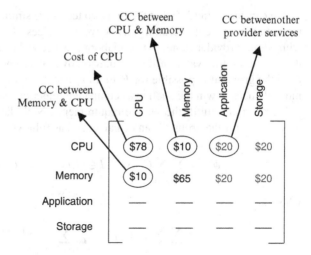

$$\phi_{rjt} = \sum_{j=1}^{n} C(s_{rjt})), \quad j \in S(P_{rt}) \tag{3.1}$$

where $C(s_{rjt})$ is the cost of supplying s_{rj} unit of services by provider r in period t. The cost $C(s_{rjt})$ for a service j can be calculated as follows from Matrix M:

$$C(s_{rjt}) = I_{C_{rjt}} e^{(\eta_{rj} \times s_{rjt})}, \quad 0 \leq s_{rjt} \leq \tilde{S}_{rjt} \tag{3.2}$$

Here,

$$I_{C_{rjt}} = {}_{j \in S(P_r)} a_{jj} + {}_{i \in S(P_r)}, \, {}_{j \in S(P_r)} a_{ij} + {}_{j \in S(P_r), k \notin S(P_r)} a_{jk} \quad \forall t \tag{3.3}$$

where, $i, j, k = 1 \ldots n$ and $i \neq j \neq k$

The first term in (3.3) is the cost of providing service j. The second term is the total collaboration cost between $S(P_r)$ services and third term refers to the total collaboration cost between services of different CPs with whom provider P_r needs to collaborate for service J. As provider P_r does not know to whom it will collaborate after winning an auction, the true cost of a_{jk} cannot be determined. Therefore, P_r may set a high collaboration cost in a_{jk} in order to avoid potential risk in collaboration phase.

Now the *BiddingFunction* of a group or group bidding price of CPs, who submit their bids collaboratively as a single bid to fulfill the service requirements completely, can be determined as follows:

Let P_r forms a group G by selecting appropriate partners where $S(P_G)$ be the set of services provided by G and $S(P_G) \subseteq R, G \subseteq P$. For any provider like $P_r \in G$, the initial cost of providing a service j can be calculated as follows

$$I_{C_{rjt}}^{G} = {}_{j \in S(P_r)} a_{jj} + {}_{i \in S(P_r)}, \, {}_{j \in S(P_r)} a_{ij} + {}_{j \in S(P_r), g \in S(P_G) \setminus S(P_r)} a_{jg} \quad \forall t \tag{3.4}$$

where, $i, j, g = 1 \ldots n$ and $i \neq j \neq g$

We can see from (3.4) that the first two terms are similar to (3.3). The third term refers to the total collaboration cost between services of other CPs inside the group with whom provider P_r needs to collaborate. Since P_r knows other group members, it can find the true value of this term. Moreover, if P_r applies any good strategy to form the group, it is possible for P_r to minimize this term. Hence, this group G has more chances to win the auction as compare to other providers who submit separate bids to partially fulfill the service requirements. Now the *BiddingFunction* or the group price for the group G can be calculated as follows:

$$\phi_{r_{jt}}^G = \sum_{r=1}^{l} \sum_{j=1}^{n} C^G(s_{rjt}), l \in G, l = 1, \ldots, G, j \in S(P_{rt}) \tag{3.5}$$

where

$$\sum_{r=1}^{l} \sum_{j=1}^{n} C^G(s_{rjt}) = \sum_{r=1}^{l} \sum_{j=1}^{n} I_{C_{rjt}}^G e^{(\eta_{rj} \times s_{rjt})} \tag{3.6}$$

subject to

$$0 \leq s_{rjt} \leq \tilde{S}_{rjt}, \quad \forall t \tag{3.7}$$

$$\sum_{r=1}^{l} \sum_{j=1}^{n} s_{rj} = R, \quad \forall t \tag{3.8}$$

3.4.2.2 Payoff Function of the Consumer

With the help of broker user generates the payoff function. During an auction, user uses the payoff function $\Omega_{\max}(R, Q)$ to internally determine the maximum payable amount that it can spend for a set of services. If the bid price of any CP is greater than the maximum payable amount $\Omega_{\max}$, it will not be accepted. In the worst case, auction terminates when the bids of all CPs are greater than $\Omega_{\max}$. In such case, user modifies its payoff function and the auctioneer reinitiates the auction with changed payoff function.

3.4.3 Winner Determination Algorithm

To solve the winner determination problem in the CACM model, we use the dynamic programming as proposed in [98]. Let $b^G(S)$ be the bid published by any group of providers G for declaring the amount of money they want to get by providing any combination of services $S \subseteq R$. The first step of winner determination is setting $b(S) = \min_{G \subseteq P} b^G(S)$. For any set of services S the lowest price in bids is selected as the price If two or more groups of providers submit same bids which are lowest among all bids, the final $b(S)$ will be judged by the time of bid acceptance.

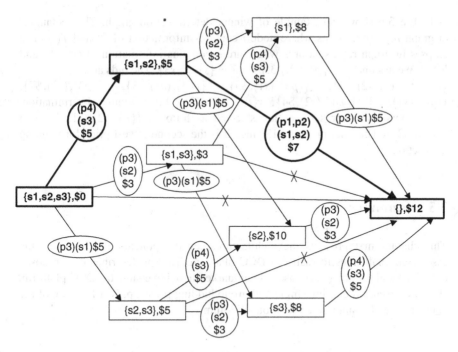

Fig. 3.6 Example of winner determination graph in our proposed auction

We define a directed graph D for winner determination. It consists of nodes $0, 1, 2, \ldots, m$ with some arcs between them for representing many states as well as conversions from one state to another. Zero is the starting node representing such a state that no service has been provided, and m is the ending node representing all services in R has been provided at that state. Any node $i(0 \leq i \leq m)$ denotes an intermediate state in which a list of services S^i is assumed as provided. For any two different nodes $i(0 \leq i \leq m)$ and $j(0 \leq j \leq m)$, if $S^i \cap S^j = S^i$, $S^j \setminus S^i$ becomes S^a, the services on the arc $a = <i, j>$ that need to be provided for switching the state from i to j. Let G^a be the group of providers of S^a. Let $w(a) = b(S^a)$ be the weight of a. Then we apply Dijkstra algorithm to calculate the shortest path from the starting node 0 to the ending node m. The length of the shortest path from 0 to m can be obtained by solving the following formula from node 1 to m.

$$f(j) = \max_{(0 \leq i < m, 0 < j \leq m)} \{w(a) + f(i)\} \quad (a = <i, j>) \tag{3.9}$$

In above formula, $f(j)$ denotes the shortest path from 0 to j. We initially set $f(0) = 0$. Then we calculate $f(1) \ldots f(m)$. that represents the shortest path from 0 to m.

For any arc a on the shortest path, providers in G^a will be selected as winners and provide S^a. As there is no "second-price" in a combinatorial auction the price paid by the winner is their bid less a discount. The discount is calculated by removing the winner from the auction and recomputing the result. The difference between the two values is the winner's discount.

Figure 3.6 shows an example of winner determination graph. The rectangles in graph represent nodes where node i contains information of S^a and $f(i)$. The ellipses in graph represent arcs where arc a contains information of G^a, S^a and $b(S^a)$. We assume $P = \{p1, p2, p3, p4\}, R = \{s1, s2, s3\}$. The bids are: $\{(p1, p2), (s1, s2), \$7\}$, $\{(p3), (s1), \$5\}$, $\{(p3), (s1), \$3\}$, $\{(p3), (s3), \$7\}$, $\{(p4), (s1), \$6\}$, $\{(p4), (s2), \$4\}$, $\{(p4), (s3), \$5\}$. After winner determination, the winners and their services are selected as follows: $\{(p1, p2), (s1, s2)\}$ and $\{(p4), (s3)\}$. The final price is \$13 since it is the second lowest price for winning the auction.

3.5   Summary

This chapter first presents the architecture, key components and identifies the associated challenges to realize a DCC platform. Then it describes the proposed combinatorial auction (CA)-based cloud market model to enable the DCC platform. Here we discuss regarding single and group bidding prices, payoff function of the consumers and winner determination algorithm.

Chapter 4
Multi-objective Optimization Model and Algorithms for Partner Selection

Abstract The Partner selection is an important decision problem in the formation of a dynamic cloud collaboration platform. Selecting suitable cloud partners to form a group will facilitate the success of collaborative cloud services. In this chapter, first, we present a promising multi-objective (MO) optimization model of partner selection considering individual information (INI) and past relationship information (PRI) with collaboration cost optimization among cloud providers in a DCC platform. Then to solve this MO optimization model, a general framework of multi-objective genetic algorithm (MOGA) that uses INI and PRI of cloud providers called MOGA-IC is presented. Finally, two algorithms called NSGA-II and SPEA2 are developed to implement MOGA-IC.

4.1 Introduction

A primary/initiator CP (pCP) identifies a business opportunity which is to be addressed by submitting a bid for a set of services for the consumer. It needs to dynamically collaborate with one or more CP partners to form a group to satisfy the consumer service requirements completely as it cannot provide all the services. We assume that each CP can provide one or at most two services and each service has one or more providers. This process of CP partner selection can be presented in Fig. 4.1. Figure 4.1 shows that the pCP ($P_{1,1}$) can provide s1 service and needs other 4 CP partners among 12 candidate CP partners to provide total five kinds of consumer service requirements (s1, s2, s3, s4 and s5).

4.2 Multi-objective Optimization Model for Cloud Partner Selection

The existing partner selection algorithms in the literature like [20,22,23,25,33,37–39,41–44,50,52,53,56,57,75,90,92,94,95,100,101] can be classified into two types

M.M. Hassan and E.-N. Huh, *Dynamic Cloud Collaboration Platform:*
A Market-Oriented Approach, SpringerBriefs in Computer Science,
DOI 10.1007/978-1-4614-5146-4_4, © The Author(s) 2013

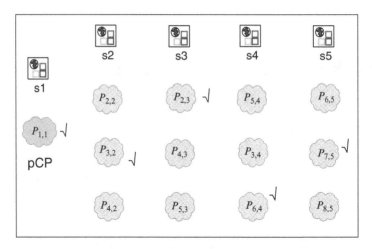

Fig. 4.1 Partner selection process for the pCP

based on the view of the relationships between tasks (to be allocated to different partners) and partners (to be selected for some specific tasks):

- One specific task with multiple candidate partners.
- Multiple tasks (organized as a whole task) in which each task has multiple candidate partners.

The first type is much simpler compared with the second type, which fully considers the relationship between the tasks instead of cost/quality optimization on a single task. Actually the relationships between tasks are very important for cloud partner selection scenario in dynamic collaboration platform. In DCC platform, cloud providers will collaboratively provide services to consumers. Currently one example can be mentioned—Salesforce.com Inc's Force.com platform now enables developers to use its cloud application development platform alongside Amazon Web Services LLC's infrastructure and storage services.

The relationship between tasks can be transformed as the collaboration between partner providers. If all tasks are allocated to one partner provider, then there will be no such collaborations. In DCC platform, multiple tasks will be allocated to multiple candidate partner providers, e.g. one partner provider for one or two tasks. So a large number of conflicts will occur since each provider must agree with services/resources contributed by other providers against a set of its own policies. Also the communications between partners must be very excessive and weighty, therefore, leading to high collaboration cost.

To select a good combination of cloud partners with less conflicts, we propose that a pCP should consider individual information (INI) and past relationship information (PRI) with collaboration cost optimization among cloud partners. In the existing partner selection approaches, the INI is mostly used, but the PRI with collaboration cost optimization among partners is overlooked. In fact, the PRI

[30, 40] with collaboration cost optimization is very important in partner selection for cloud providers in dynamic collaboration environment. The success of past relation with less collaboration costs between participating cloud providers may reduce uncertainty and conflicts, short adaptation duration, and also help to the performance promotion.

4.2.1   Optimization Goals of Cloud Partner Selection

The factors to be considered during cloud partner selection include quality, cost (costs of tasks and cost for collaborations between tasks), number of past relations, time etc. and the solution of partner selection problem (PSP) is to realize total optimization of such factors. According to the properties of these goals, they may be divided into three types:

- Individual goals, e.g. quality, cost for a service/task. They are related to the capacity of selected partners only.
- Collaboration goals, e.g. number of past relation between partners, cost of collaborations between services/tasks. They depend on the collaborations between tasks and other partners.
- Constraint goals, e.g. time, capacity. They not only depend on the execution of each task but are also related to the collaboration time between partners.

The past relationship information include number of projects/auctions accomplished/won by other providers among themselves and with the pCP. The collaboration cost include network establishment cost (after one provider finishes a task, it should be transported to another partner provider for further processing e.g. virtual machine (VM) running locally at a certain provider must be migrated to another provider for further processing), information transmission cost (partners have to exchange some information like QoS requirements, SLA, files, security measure etc. for further task processing) and capital flow cost (one partner provider has to pay some expense to another partner provider e.g. VM running and monitoring cost etc.).

The following notations are used for the mathematical formulation of cloud partner selection problem:

4.2.2   Problem Formulation of Cloud Partner Selection

The optimal solution of cloud partner selection is to select a group of cloud partners v from m candidates who collaboratively win auctions many times with less collaboration costs (maximizing past relationship performance values) and making the individual price the lowest and quality value of service the highest. In the most situations, it is impossible that there is a candidate provider group that can make

$R = \{R_{jt} \mid j = 1 \ldots n\}$	A set of service requirements of consumer in period t ($t = 1 \ldots T$)
$P = \{P_r \mid r = 1 \ldots m\}$	A set of cloud providers who participate as bidders
V_t	Number of cloud providers to be selected in period t($V \subset P$)
P_{rjt}	A cloud provider r who can provide service j in period t
$\tilde{S}_{rjt}$	Capacity of cloud provider r for service j in period t
SP_{rjt}	Unit price of service j of provider r in period t
NS_{rjt}	Usage of number of units of service j of provider r in period t
$CC_{rj,xi}$	Total collaboration cost between provider r of service j and x of service i
$MC_{rj\leftrightarrow xi}$	Unit cost of migration of service j and i between providers r and x (e.g. VM migration between partners)
$IT_{rj\leftrightarrow xi}$	Unit cost of information transmission of service j and i between providers r and x
$CF_{rj\leftrightarrow xi}$	Unit cost of capital flow service j and i between providers r and x
ϕ_{rjt}	The price of CP r for providing service j independently in period t
Q_{rjt}	The quality value for service j of CP r in period t(qualitative information can be expressed by the assessment values from 1 to 10 (1: very bad, 10: very good))
$W_{rj,xi}$	The value of past relationship experience (i.e. number of times collaboratively wins auctions or did projects) between a provider r for service j and other provider x for service i where ($r,x = 1 \ldots m; i,j = 1 \ldots n; i \neq j$)
U_{rjt}	A decision vector of partner selection in period t

all the goals optimized. So to solve the PSP of a pCP using the INI and PRI, a MO optimization model to minimize total price and maximize service quality and total collaborative past relationship (PR) performance with collaboration cost optimization values can be expressed mathematically as follows:

$$\text{Minimize Obj\_1} = \sum_{j=1}^{n} \sum_{r=1}^{m} \sum_{t=1}^{T} SP_{rjt} NS_{rjt} U_{rjt} \tag{4.1}$$

$$\text{Maximize Obj\_2} = \sum_{j=1}^{n} \sum_{r=1}^{m} \sum_{t=1}^{T} Q_{rjt} U_{rjt} \tag{4.2}$$

$$\text{Maximize Obj\_3} = \sum_{\substack{i,j=1 \\ i \neq j}}^{n} \sum_{\substack{r,x=1 \\ r \neq x}}^{m} \sum_{t=1}^{T} W_{rj,xi} U_{rjt} U_{xit} + \sum_{\substack{i,j=1 \\ i \neq j}}^{n} \sum_{\substack{r,x=1 \\ r \neq x}}^{m} \sum_{t=1}^{T} CC_{rj,xi} U_{rjt} U_{xit} \tag{4.3}$$

subject to

$$U_{rjt} = \begin{cases} 1 \text{ if choose } P_{rjt} \\ 0 \text{ otherwise} \end{cases} \tag{4.4}$$

$$U_{rjt} U_{xi} = \begin{cases} 1 \text{ if choose } P_{rjt} \text{ and } P_{xit} \\ 0 \text{ otherwise} \end{cases} \tag{4.5}$$

$$\sum_{j=1}^{n}\sum_{r=1}^{m}\sum_{t=1}^{T}U_{rjt}=V_t \quad (r=1\ldots..m, j=1\ldots.n, t=1\ldots T) \qquad (4.6)$$

$$\sum_{\substack{i,j=1\\i\neq j}}^{n}\sum_{\substack{r,x=1\\r\neq x}}^{m}\sum_{t=1}^{T}CC_{rj,xi}U_{rjt}U_{xit}=\sum_{\substack{i,j=1\\i\neq j}}^{n}\sum_{\substack{r,x=1\\r\neq x}}^{m}\sum_{t=1}^{T}\left(\frac{\log_2\left[\frac{Max(MC_{rj\leftrightarrow xi}IT_{rj\leftrightarrow xi}CF_{rj\leftrightarrow xi})}{MC_{rj\leftrightarrow xi}IT_{rj\leftrightarrow xi}CF_{rj\leftrightarrow xi}U_{rjt}U_{xit}+1}\right]}{\log_2\{Max(MC_{rj\leftrightarrow xi}IT_{rj\leftrightarrow xi}CF_{rj\leftrightarrow xi})\}}\right)\cdot$$

$$(4.7)$$

$$\sum_{j=1}^{n}\sum_{r=1}^{m}\sum_{t=1}^{T}\tilde{S}_{rjt}U_{rjt}\geq R_{jt} \qquad \forall r,j,t \qquad (4.8)$$

For model (4.1)–(4.3), its solution space is mainly a function of parameters m and v. Let F denote the number of solutions in the solution space, there exists $F = C_m^v = \frac{m!}{(m-v)!v!}$. Because of the property of combination number, there is $C_m^v = C_m^{m-v}$, such that $v \leq \frac{m}{2}$ or $m-v \leq \frac{m}{2}$. In the case that v is much smaller than m, i.e., $v \ll m$,, F could be approximately processed as follows:

$$F = \frac{m!}{(m-v)!v!} = \frac{m(m-1)\ldots..(m-v+1)}{v!} \approx m(m-1)\ldots\ldots.(m-v+1) \approx m^v$$

$$(4.9)$$

According to the above analysis, the solution space will approximately exponentially grow with increasing v. Meanwhile, since model (4.1)–(4.3) is NP-hard, the solving for model is not easy. For the problem with small scale, i.e., m and v is very small, traditional enumeration method [20] is capable. However, for the large scale problem, an intelligent optimization algorithm is required.

4.3   General Multi-objective Genetic Algorithm Framework for Cloud Partner Selection

For solving PSP of cloud providers, we develop a general framework of multi-objective genetic algorithm (MOGA) that uses individual information (INI) and past collaborative relationship information (PRI) with collaboration cost optimization among cloud providers called MOGA-IC as follows:

Common framework for MOGA-IC:

Step 1: Initialize the population P.
Step 2: Conduct a selection operation to select elite individuals based on fitness function (evaluating multi-objective functions) from P and store their data in external set E (optional and not for non-elitist MOGAs).

Step 3: Create a mating pool using either P or E, or both.
Step 4: Apply crossover and mutation operators.
Step 5: Evaluate individuals.
Step 6: Conduct reproduction based on the pool to create the next generation of P.
Step 7: Combine P and E.
Step 8: If termination criteria are not satisfied, return to Step 2.

When solving PSP for CPs using MOGA technique, one important issue needs to be addressed: how to find an appropriate diversity preservation mechanism in selection operators to enhance the yield of Pareto optimal solutions during optimization, particularly for the CP PSPs having multiple conflicting objectives. So we apply two popular MOGAs- the non-dominated sorting genetic algorithm (NSGA-II) [32] and the strength Pareto evolutionary genetic algorithm (SPEA2) [102] to the general framework of MOGA-IC, both of which include an excellent mechanism for preserving population diversity in the selection operators.

4.3.1 Solution Encoding and Generating Initial Population

Natural number encoding is adopted to represent the chromosome of individual. A chromosome of an individual is an ordered list of CPs. Let $y = [y_1, y_2, \ldots, y_j \ldots \ldots y_n](j = 1, 2, \ldots n), y_j$ be a gene of the chromosome, with its value between 1 and m (for service j, there are m CPs for a response). If $m = 50$ and $n = 5$, there may be 10 CPs that can provide each service j. Thus a total of 10^5 possible solutions is available. In this way, the initial populations are generated.

4.3.2 Crossover and Mutation Operator

The crossover operator is a very important operator in MOGA. The main idea behind crossover is that a combination between segments of two individuals might yield a new individual which benefits from both parents advantages. The crossover operator gets two individuals as its input, and outputs two new individuals (i.e. the offspring). Two-point crossover is employed in MOGA-IC.

The role of the mutation operator [32] is to lower the probability of converging at a local optimum. In the case of mutation, one provider is randomly changed for any service. Evidently, this method only changes the order of the genes but dose not change the number of the genes. Thus, no infeasible solutions are produced.

4.3.3 Fitness Calculation

In the single objective decision making model, the fitness function is usually the objective function. However, in the multi-objective models, the multi-objective functions should be considered when calculating the fitness values. In the case of cloud partner selection problem, we have three multi-objective functions- minimization of the price of service while maximization of past collaborative relationship performance and service quality values. We use non-dominated sorting approach to calculate the fitness value of individual. Any two individuals are selected and their corresponding fitness values are compared according to the dominating-relationships. Before using non-dominated sorting approach objective functions must be normalized using the (4.10) because they have different measure units.

$$f'_i = \frac{f_i - f_i^{min}}{f_i^{max} - f_i^{min}}, i = 1.2., \ldots, n \tag{4.10}$$

4.3.4 Selection Algorithms

The selection algorithm is an important component when solving PSP for CPs using MOGA techniques, since it steers the search direction when searching for optimal solutions. In MO optimization using a GA, elitism with a diversity preservation mechanism is often preferred to obtain the generally high quality of the obtained optimal solutions. The selection method used in this paper implements NSGA-II and SPEA2 algorithms. Based on different principles, they both have excellent mechanism for the preservation of diversity.

4.3.4.1 MOGA-IC with NSGA-II

NSGA-II is a fast, elite MOGA proposed by Deb et al. [8]. The steps are described as follows:

Algorithm: MOGA-IC with NSGA-II

Step 1 Initialize the input parameters which contain the number of requirements (R), providers (m) and maximum genetic generations (G), population size (N), crossover probability (P_c) and mutation probability (P_m).

Step 2 Generate the initial parent population $P_t, (t = 0)$ of size N_p.

Step 3 Apply binary tournament selection strategy to the current population, and generate the offspring population O_t of size $N_o = N_P$ with the predetermined P_c and P_m .

Step 4 Set $S_t = P_t \cup O_t$, apply a non-dominated sorting algorithm based on fitness function (evaluating multi-objective functions) and identify different fronts $F_1, F_2 \ldots F_a$.

Step 5 If the stop criterion ($t > G$) is satisfied, stop and return the individuals (solutions) in population P_t and their corresponding objective values as the Pareto-(approximate) optimal solutions and Pareto-optimal fronts.

Step 6 Set new population $P_{t+1} = 0$. Set counter $i = 1$. Until $|P_{t+1}| + |F_i| \leq N$ set
        $P_{t+1} = P_{t+1} \cup F_i$ and $i = i + 1$.

Step 7 Perform the crowding-sort procedure and include the most widely spread
        $(N - |P_{t+1}|)$ solutions found using the crowding distance values in sorted F
        in P_{t+1}.

Step 8 Apply binary tournament selection, crossover and mutation operators to
        P_{t+1} to create offspring population O_{t+1}.

Step 9 Set $t = t + 1$, then return to Step 4.

4.3.4.2 MOGA-IC with SPEA2

SPEA2 [9] is a very effective algorithm that uses an external list to store non-dominated solutions discovered in the course of searching. It is an excellent example for the use of external populations. The steps are described as follows:

Algorithm: MOGA-IC with SPEA2

Step 1:   Generate a random population P_0 of size N_P. Set $t = 0$ and an empty
           external archive E_0 of size N_E.

Step 2:   Calculate the fitness of each solution x in $P_t \cup E_t$ as follows:

Step 2.1: Calculate the raw fitness as $R(x,t) = \sum_{y \in P_t \cup E_t, y \succ x} S(y,t)$ where $S(y,t)$ is
           the number of solutions in $P_t \cup E_t$ dominated by solution y.

Step 2.2: Calculate the density as $D(x,t) = (\sigma_x^k + 2)^{-1}$, where σ_x^k is the distance
           between solution x and its kth nearest neighbor, where $k = \sqrt{N_p + N_E}$.

Step 2.3: Assign a fitness value as $F(x,t) = R(x,t) + D(x,t)$.

Step 3:   Copy all non-dominated solutions in $P_t \cup E_t$ to E_{t+1}. Now, two cases
           may arise. Case 1: If $|E_{t+1}| > N_E$, then truncate $|E_{t+1}| - N_E$ solutions by
           iteratively removing solutions that have maximum σ^k distances. Break
           any tie by examining σ^l for $l = k - 1, \ldots, 1$ sequentially. Case 2: If
           $|E_{t+1}| \leq N_E$, copy the best $N_E - |E_{t+1}|$ dominated solutions according to
           their fitness values from $P_t \cup E_t$ to $E_t + 1$

Step 4:   If the stopping criterion is satisfied, stop and copy the non-dominated
           solutions in E_{t+1}.

Step 5:   Select the parent from E_{t+1} using binary tournament selection with
           replacement.

Step 6:   Apply the crossover and mutation operator to the parents to create N
           offspring solutions. Copy offspring to $P_{t+1}, t = t + 1$, then return to Step 2.

4.4 Summary

In this chapter, first, we discuss regarding the optimization goals for cloud partner selection in a DC platform and present a new promising MO optimization model of partner selection considering INI and PRI with collaboration cost optimization among cloud providers. Second, a general framework of MOGA that uses INI and

PRI with collaboration cost optimization among cloud providers called MOGA-IC is presented to solve this MO optimization model. Finally, two algorithms called NSGA-II and SPEA2 are developed to implement MOGA-IC to find an appropriate diversity preservation mechanism in selection operators to enhance the yield of Pareto optimal solutions during optimization, particularly for the CP PSPs having multiple conflicting objectives.

Chapter 5
Experimental Results and Analysis

Abstract In this chapter, we present our evaluation methodology and simulation results of the utility of dynamic cloud collaboration (DCC), MOGA-IC for CP partner selection and proposed CACM model. First, we compare DCC platform with existing cloud federation model. Then, we present the comparison of the proposed CACM model with the existing CA model in terms of economic efficiency. Next, we present a simulation example of a partner selection problem (PSP) for a pCP in the CACM model. It is used to illustrate the proposed MOGA-IC method. NSGA-II and SPEA2 are utilized to develop the MOGA-IC. Further simulation examples are conducted to pinpoint the most viable approach (NSGA-II or SPEA2) for MOGA-IC. Moreover, we implement the existing MOGA that uses only INI called MOGA-I for CP partner selection and analyze its performance with MOGA-IC in the proposed CACM model. Finally, numerical results are presented to demonstrate the utility of the price minimization algorithm among cloud providers in the CACM model.

5.1 Introduction

The dynamic cloud collaboration (DCC) platform offers efficient collaborative cloud services with high availability, transparency, and improved performance without requiring consumers to build or manage complex infrastructure themselves. Consumers interact with the DCC system in a limited number of ways and have little experience of the associated complex technologies. The responsibility of ensuring high-performance cloud service delivery is largely on the DCC system itself. Therefore, a comprehensive analysis of the impact of several system parameters on utility is important to improve the system's service ability.

The simulation methodology realizes a privileged provider model, with the primary cloud provider (pCP) having the authoritative right over the resources it has acquired, which are delegated rights for the collaborators' physical resources. The pCP recognizes a good business opportunity in the market and then dynamically

M.M. Hassan and E.-N. Huh, *Dynamic Cloud Collaboration Platform:*
A Market-Oriented Approach, SpringerBriefs in Computer Science,
DOI 10.1007/978-1-4614-5146-4_5, © The Author(s) 2013

collaborate with suitable partner CPs to form a group and submits their group bid as a single bid in the CACM in order to provide a set of complementary services to consumers. If this group wins the auction for a set of services, they provide composite or collaborative services to consumers. Consumers interact transparently with the DCC platform by requesting services through a service catalog of the pCP. The CPs offer capabilities/services to consumers with a full consumption specification formalized as a standard SLA. The requested service requirements (single, multiple or collaborative cloud services) are served either directly by the pCP or by any collaborating CP within the DCC.

5.2 Simulation Environment and Parameters

One of the main challenges to evaluate the DCC platform, the partner selection problem and the CACM model is the lack of real-world input data. So we conduct the experiments using synthetic data. For the experiment of measuring the utility of a DCC platform, the workloads have been generated using Lublin's model [62] from parallel workload archives (http://www.cs.huji.ac.il/labs/parallel/workload/) because trace data of cloud applications are currently not released and shared by any commercial cloud service providers. However, for a scientific research paper, it is extremely important to have publicly accessible trace data so that the experiments can be reproducible by other researchers. Moreover, this paper focuses on studying the application requirements of users in the context of High-Performance Computing (HPC). Hence, the Parallel Workload Archive meets our objective by providing the necessary characteristics of real parallel applications collected from supercomputing centers.

The Lublin model has been configured to generate two-month long workloads of typeless requests (i.e. no distinction is made between batch and interactive requests). Unfortunately, since the Parallel Workload Archive is not based on paying users in utility computing environments, it is possible that the trace pattern of these archived workloads will be different from those with paying users. Furthermore, the Lublin model only provides the inter-arrival times of reservation requests, the number of nodes to be reserved and the duration to be reserved. Service requirements are not available in this model. Hence, we use uniform distribution to assign service requirements synthetically to Lublin model. To generate price of cloud services in the CACM model, we follow the pricing idea for Cl41oud services presented in [7, 97]. If any provider has more collaboration experience with other providers, the collaboration cost (CC) can be minimized. We use the following formula to calculate the CC between any provider P_{rj} and P_{xi}:

$$CC_{rj,xi} = MinCC_{rj,xi} + (MaxCC_{rj,xi} - MinCC_{rj,xi}) \times \frac{1}{e^{W_{rj,xi}}} \qquad (5.1)$$

Table 5.1 Basic parameter values used for simulation

Parameters	Description	Value
R	Service requirements	5–10
P	Number of cloud providers	10–100
$S(P_{rt})$	Service per provider	1–2
SP_{rjt}	Unit price of service	0.1–0.3\$/h
$MinCC_{rj,xi}$	Minimum CC between services	10\$
$MaxCC_{rj,xi}$	Maximum CC between services	20\$
Q_{rjt}	Quality value	1–10
$W_{rj,xi}$	Past relationship value	0–10

where
$MinCC_{rj,xi}$= the minimum CC between services
$MaxCC_{rj,xi}$= the maximum CC between services
$W_{rj,xi}$ = the number of collaboration experiences between P_{rj} and P_{xi}. If it is zero, the highest CC is set between providers. Thus the final price of services is generated for each provider, and it is varied based on CC in different auctions.

For partner selection problem, we provide simulation examples using synthetic data. We have used Visual C++ language to conduct repeatable and controlled simulations. The experiments were conducted on a Intel Pentium Core 2 duo having the following configuration: CPU of 2.33 GHz with 2 MB L2 cache, 1.98 GB of RAM and 250 GB of storage. The basic parameters used for all simulations are shown in Table 5.1.

5.3 Simulation Results

5.3.1 Utility of Dynamic Cloud Collaboration

To show the usage benefits of dynamic cloud collaboration, we compare its performance with non-collaborated approach and existing static collaboration approach. We use two metrics to measure its utility—percentage of service rejection and overall resource utilization.

In order to generate different workloads, we modify two parameters of Lublin99's model, one at a time. To generate varying size of workloads, we vary the parameter called *umed* in Lublin99's model from 1.5 to 3.5. The larger the value of *umed*, the smaller the requests become in terms of numbers of VMs required and consequently, result in lighter loads. The second parameter changed in the experiments affects the inter-arrival time of requests at rush hours. The inter-arrival rate of jobs is modified by setting the β of the gamma distribution (hereafter termed *barr*), which we vary from 0.46 to 0.60. As the values for *barr* increase, the inter-arrival time of requests also increases.

Table 5.2 Capacity of 10 cloud providers

Cloud providers	CPU units	Memory units	Storage units	Network units
1	6,000	2,000	4,000	1,000
2	5,000	3,000	3,000	1,000
3	4,060	2,500	3,050	1,000
4	3,048	2,056	2,050	1,000
5	2,024	3,128	3,500	1,000
6	3,200	4,500	2,000	1,000
7	5,011	3,000	1,500	1,000
8	6,050	2,500	2,050	1,000
9	1,500	1,500	1,050	1,000
10	1,250	1,300	3,500	1,000

We run our experiments with one provider as primary cloud provider and nine providers as collaborators. Results are averaged over ten simulation runs. We consider static collaboration consists of five cloud providers whereas dynamic cloud collaboration can have any number (within nine providers) of cloud providers based on service requirements. We generate the capacity of Ten cloud providers as shown in Table 5.2 following the model of [56].

We also develop a metric to measure overall resource utilization. Let us consider there are $N(CP)$ cloud providers and each provider has total $N(P)$ physical machines. Also assume that a combinatorial service requirement containing $N(t)$ tasks with QoS specifications need to be allocated. For the non-collaborated approach, a cloud provider will try to allocate $N(t)$ tasks to its physical machines. For the static and dynamic collaboration cases, $N(t)$ tasks can be shared among collaborative partners.

For a physical machine p_j $(1 <= j <= N(p))$, let oc_j, om_j, os_j and ob_j be the original resource conditions of its CPU, memory, storage and network bandwidth. If a task t_i is allocated to a physical machine p_j (denoted as $t_i \in p_j$), it should occupy a part of resource on p_j. Let fc_j, fm_j, fs_j and fb_j be the percentage of idle CPU, memory, hard-disk and network bandwidth resource on p_j. so the overall resource utilization U can be calculated as follows:

$$U = \lambda_c \left[\sum_{i=1}^{N(CP)} \sum_{j=1}^{N(p)} (1 - fc_{ij}) \times oc_{ij} \right] + \lambda_m \left[\sum_{i=1}^{N(CP)} \sum_{j=1}^{N(p)} (1 - fm_{ij}) \times om_{ij} \right]$$

$$+ \lambda_s \left[\sum_{i=1}^{N(CP)} \sum_{j=1}^{N(p)} (1 - fs_{ij}) \times os_{ij} \right] + \lambda_b \left[\sum_{i=1}^{N(CP)} \sum_{j=1}^{N(p)} (1 - fb_{ij}) \times ob_{ij} \right] \quad (5.2)$$

where $\lambda_c, \lambda_m, \lambda_s$ and λ_b denote the pricing scheme for CPU capacity, memory space, storage space and network bandwidth.

5.3.1.1 Performance of DCC in Terms of Service Rejection

We first present our experimental results of the performance of DCC in terms of service rejection. We vary the *umed* and *barr* values and the results are shown in Fig. 5.1. We can see from Fig. 5.1 that the lower values of umed and barr result in higher loads and thus rejection rate becomes high. Still under heavy workload, DCC rejects less service or tasks as compared to the non-collaborated and existing static collaboration approaches. For example, when the *umed* value is 1.5, non-collaborated approach, static collaboration approach and DCC approach reject the service request of 62%, 32% and 27% respectively. In addition, in case of small inter-arrival period (e.g. barr = 0.46), DCC, static collaboration and non-collaborated approach reject 27%, 35% and 65% of the service request respectively. Thus, DCC outperforms the static collaboration and non-collaborated approaches. The reason is that based on the service request in a DCC platform, suitable partners can be selected who can fulfill the service request while in case of static approaches, fixed or pre-existing partners can only fulfill the service requests. Hence, rejection rate becomes very high. Also service heterogeneity makes static collaboration to reject many jobs since it cannot always predict the incoming service behaviour and requirements.

5.3.1.2 Performance of DCC in Terms of Resource Utilization

In this experiment, we measure overall resource utilization (e.g. cpu, memory, storage and network) based on (5.2) in the DCC, static and non-collaborated environments by varying *umed* and *barr* values. Figure 5.2 shows the result of the experiment under Min–Min heuristic based scheduling [30]. We can see from Fig. 5.2 that in heavy and light workload scenarios (both *umed* and *barr* case), the percentage of the overall resource utilization in DCC is always higher as compared to that of in the non-collaborated approach and static collaboration approach. The reason is that in a DCC environment, computing intensive tasks are executed by those CPs, which have high-performance CPU while storage intensive tasks are run by those CPs, which have large size of storage space, which means in a DCC platform, suitable partners are selected based on service requirements. This can improve the overall performance for combinatorial tasks as well as utilization of resources.

5.3.2 Economic Efficiency of CACM Model as Compared to the Exiting CA Model

The proposed auction policy in the CACM model allows any CP to collaborate dynamically with appropriate partner CPs to form groups before joining the auction

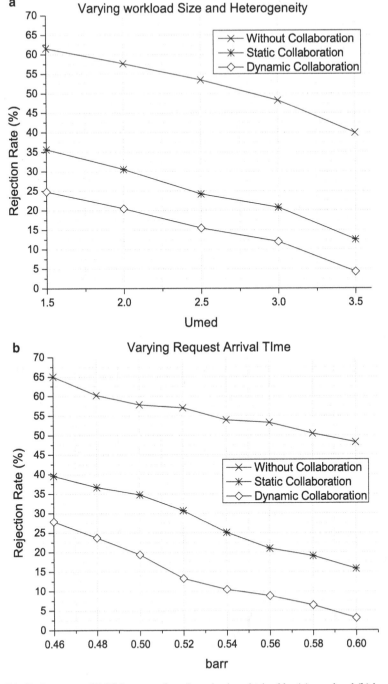

Fig. 5.1 Performance of DCC in terms of service rejection obtained by (**a**) umed and (**b**) barr

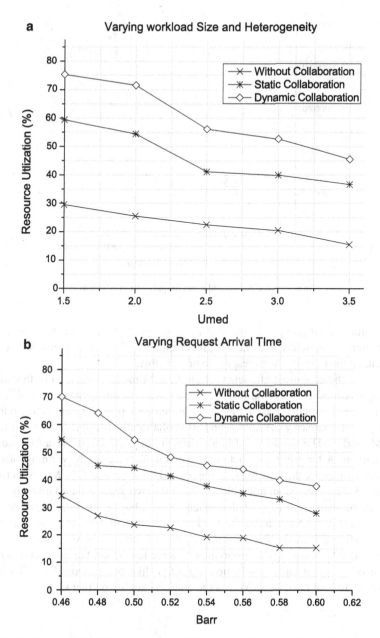

Fig. 5.2 Performance of DCC in terms of overall resource utilization obtained by (**a**) umed and (**b**) barr

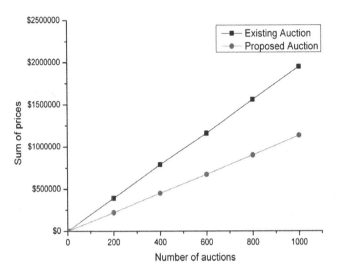

Fig. 5.3 Economic efficiency of CACM model as compared to the existing CA model

and to publish their group bids as a single bid in order to fulfill the consumer service requirements completely. This new approach enables CPs to minimize conflicts and to calculate the true CCs with respect to one another.

To show the economic efficiency of the CACM model compared to that of the existing CA model, in our simulation, 1,000 auctions are generated for different consumer requirements. Based on those requirements, providers separately publish their bids to the existing CA market and also collaboratively publish their bids to the CACM market. The winners and final prices are determined by the auctioneers in both markets. After every 200 auctions, we count the prices of the winning bids that were determined by the existing CA market and the proposed CACM, respectively. Figure 5.3 shows the economic efficiencies of the two auction-based markets.

It can be seen from Fig. 5.3 that when the number of auctions' increases, the CACM auction model reduces the total service price to consumers as compared to the existing CA model for the same number of service requirements. The main reason is that CCs among the group members are lower, and the total service price is reduced. A good partner selection algorithm like our proposed MOGA-IC is required to reduce the CCs as well as the conflicts among partner CPs.

5.3.3   Appropriate Approach to Develop the MOGA-IC

In this section, we present three simulation examples of PSP for a pCP in the CACM model. Table 5.3 shows the three simulation examples with MOGA-IC parameters for partner selection problem in the CACM model. For each simulation example,

Table 5.3 The three simulation examples with MOGA-IC parameters

Simulation examples	No. of providers m	Service req. R	Population size N/E	Maximum genetic generation G	Crossover probability P_c	Mutation probability P_m
1	35	5	50	20	0.9	0.1
2	100	5	100	50	0.9	0.1
3	100	5	100	100	0.9	0.1

Table 5.4 The normalized individual information of pCP and other candidate CPs

Service no.	Provider no.	Price of service	Quality value of service
2	28	0.17	0.99
2	10	038	0.88
3	10	0.99	0.88
3	15	0.81	0.3
3	6	0.66	0.01
3	32	0.07	0.65
3	14	0.55	0.23
3	20	0.88	0.72
4	9	0.00	0.54
4	33	0.17	0.4
4	18	0.84	0.62
4	17	0.89	0.02
4	34	0.5	0.66
4	26	0.57	0.00
7	1	0.4	1
8	2	0.83	0.19
8	21	0.73	0.48
8	11	0.63	0.06
8	23	0.94	0.22
8	19	0.81	0.63
8	32	0.88	0.82

MOGA-IC is developed based on NSGA-II and SPEA2. Furthermore, in each simulation example, two individual informations (price and quality of services) and one past collaborative relationship information (number of auctions collaboratively won by other providers among themselves and also with pCP) of candidate CPs are considered. Both pieces of information are presented in Tables 5.4 and 5.5 in normalized forms for the first simulation example. For normalization, the method proposed by Hwang and Yoon [93] is utilized. We can see that 21 total CPs are found from 35 candidate CPs who can provide five randomly generated consumer service requirements. We assume that provider number **1** is the pCP which can provide the service number **7**. The number of generation's G in the first simulation example is set to 20 since the example search space is quite small.

Table 5.5 The normalized past relationship information (PRI) of pCP and other candidate CPs

	P 28, 2	P 10, 2	P 10, 3	P 15, 3	P 6, 3	P 32, 3	P 14, 3	P 20, 3	P 9, 4	P 33, 4	P 18, 4	P 17, 4	P 34, 4	P 26, 4	P 1, 7	P 2, 8	P 21, 8	P 11, 8	P 23, 8	P 19, 8	P 32, 8
P 28, 2	–	–	0.51	0.5	0.79	0.25	0.79	0.14	0.69	0.66	0.24	0.54	0.18	0.3	0.29	0.28	0.36	0.42	0.96	0.97	0.72
P 10, 2	–	–	0.28	0.62	0.7	0.52	0.51	0.48	0.31	0.02	0.81	0.22	0.74	0.94	0.79	0.17	0.4	0.03	0.4	0.39	0.77
P 10, 3	–	–	–	–	–	–	–	–	0.49	0.67	0.28	0.13	0.41	0.63	0.93	0.66	0.17	0.01	0.70	0.26	0.96
P 15, 3	–	–	–	–	–	–	–	–	0.65	0.57	0.04	0.98	0.18	0.08	0.13	0.83	0.66	0.84	0.63	0.20	0.23
P 6, 3	–	–	–	–	–	–	–	–	0.48	0.38	0.28	0.18	0.38	0.27	0.81	0.11	0.77	0.79	1.0	0.29	0.96
P 32, 3	–	–	–	–	–	–	–	–	0.39	0.04	0.09	0.84	0.87	0.35	0.79	0.16	0.43	0.87	0.11	0.80	0.25
P 14, 3	–	–	–	–	–	–	–	–	0.68	0.54	0.29	0.32	0.21	0.44	0.85	0.09	0.18	0.666	0.19	0.52	0.74
P 20, 3	–	–	–	–	–	–	–	–	0.05	0.41	0.81	0.33	0.04	0.01	0.90	0.28	0.01	0.03	0.67	0.82	0.01
P 9, 4	–	–	–	–	–	–	–	–	–	–	–	–	–	–	0.17	0.51	0.56	0.26	0.07	0.56	0.93
P 33, 4	–	–	–	–	–	–	–	–	–	–	–	–	–	–	0.07	0.22	0.50	0.59	0.87	0.16	0.77
P 18, 4	–	–	–	–	–	–	–	–	–	–	–	–	–	–	0.68	0.57	0.41	0.91	0.88	0.04	0.87
P 17, 4	–	–	–	–	–	–	–	–	–	–	–	–	–	–	0.64	0.21	0.50	0.04	0.73	0.02	0.67
P 34, 4	–	–	–	–	–	–	–	–	–	–	–	–	–	–	0.51	0.09	0.11	0.13	0.75	0.49	0.77
P 26, 4	–	–	–	–	–	–	–	–	–	–	–	–	–	–	0.27	0.32	0.68	0.57	0.31	0.07	0.05
P 1, 7	–	–	–	–	–	–	–	–	–	–	–	–	–	–	–	0.22	1.0	0.8	0.47	0.58	0.96

Table 5.6 Pareto-optimal solutions of MOGA-IC with NSGA-II for example 1

Pareto-optimal solutions					Optimal objective function values		
$y =$ y_7	y_4	y_3	y_2	y_8	Obj_1	Obj_2	Obj_3
1	18	6	10	32	3.16	3.32	7.63
1	18	10	10	32	3.49	4.2	7.33
1	34	10	28	32	2.94	4.35	6.24
1	9	32	28	21	1.37	3.66	4.93
1	9	10	28	32	2.44	4.23	6.65
1	9	32	28	19	1.45	3.81	5.49
1	9	14	28	32	2.00	3.58	6.82
1	34	32	10	32	2.23	4.01	6.97
1	18	10	28	32	3.28	4.31	6.44
1	9	32	10	32	1.73	3.89	5.88
1	34	32	28	32	2.02	4.12	5.59
1	34	6	10	32	2.82	3.36	7.39
1	34	32	10	19	2.16	3.82	6.48
1	34	10	10	32	3.15	4.24	7.12
1	9	32	28	32	1.52	4.00	5.44
1	18	14	10	32	3.05	3.55	7.27

Table 5.7 Pareto-optimal solutions of MOGA-IC with SPEA2 for example 1

Pareto-optimal solutions					Optimal objective function values		
$y =$ y_7	y_4	y_3	y_2	y_8	Obj_1	Obj_2	Obj_3
1	9	14	28	19	1.93	3.39	6.10
1	9	32	28	32	1.52	4.00	5.44
1	9	32	28	19	1.45	3.81	5.49
1	9	32	28	11	1.27	3.24	4.93
1	9	14	28	32	2.00	3.58	6.82
1	9	10	28	32	2.44	4.23	6.65
1	34	32	28	19	1.95	3.93	5.73
1	34	32	28	32	2.02	4.12	5.59
1	9	32	10	32	1.73	3.89	5.88
1	34	10	10	32	3.15	4.24	7.12
1	34	32	10	19	2.16	3.82	6.48
1	34	32	10	21	2.08	3.67	6.16
1	9	32	28	21	1.37	3.66	4.93
1	18	10	10	32	3.49	4.2	7.33
1	34	10	28	32	2.94	4.35	6.24
1	34	6	10	32	2.82	3.36	7.39

In solving the first simulation example problem of CP partner selection, the best Pareto front among the ten trials of 20 generations is selected as the final solution. The 16 pareto-optimal solutions of the first front of MOGA-IC with NSGA-II and SEPA2 for simulation example 1 are presented in Tables 5.6 and 5.7, respectively. Graphical representations are shown in Fig. 5.4.

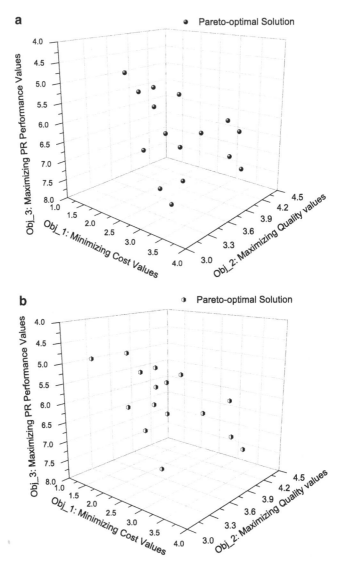

Fig. 5.4 Pareto-optimal solutions of MOGA-IC for simulation example 1 (N/E = 50 and G = 20) obtained by (**a**) NSGA-II and (**b**) SPEA2

From Fig. 5.4, it is difficult to compare the performances of MOGA-IC with NSGA-II and SPEA2 since the solution space is quite small. We conduct further performance tests of MOGA-IC with NSGA-II and SPEA2 using simulation examples 2 and 3. Figures 5.5 and 5.6 show plots of Pareto-optimal solution sets of the first fronts obtained by MOGA-IC using NSGA-II and SPEA2 when solving

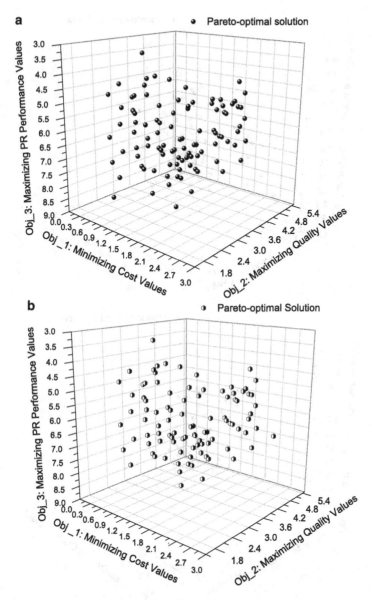

Fig. 5.5 Pareto-optimal solutions of MOGA-IC for simulation example 1 (N/E = 100 and G = 50) obtained by (**a**) NSGA-II and (**b**) SPEA2

the simulation examples 2 and 3, respectively. Here, we only provide graphical representations of the solutions for both the algorithms as the input data tables are very large.

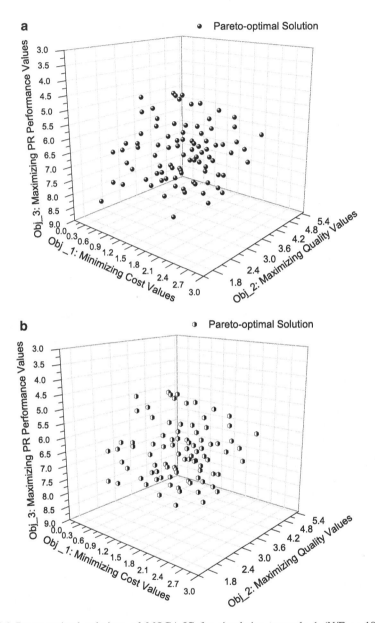

Fig. 5.6 Pareto-optimal solutions of MOGA-IC for simulation example 1 (N/E = 100 and G = 100) obtained by (**a**) NSGA-II and (**b**) SPEA2

In Figs. 5.5 and 5.6, the Pareto fronts obtained using MOGA-IC with SPEA2 are dominated by MOGA-IC with NSGA-II solutions. To verify the inferior performance of SPEA2, Figs. 5.7 and 5.8 show the average optimized values of

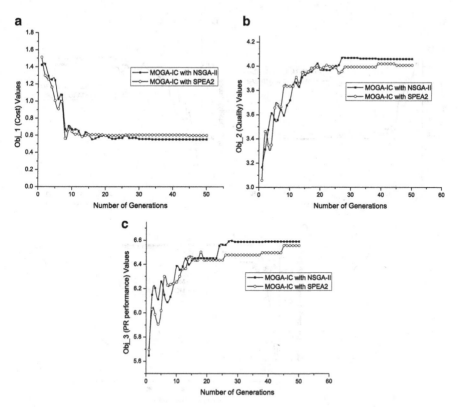

Fig. 5.7 Average optimized values of different objective functions in the first front of MOGA-IC with NSGA-II and SPEA2 for 50 generations. (**a**) Optimized avg. values of Obj_1 (cost) in NSGA-II and SPEA2. (**b**) Optimized avg. values of Obj_2 (quality) in NSGA-II and SPEA2 (**c**) Optimized avg. values of Obj_3 (PR performance) in NSGA-II and SPEA2

three objective functions in the first fronts during 50 and 100 generations using MOGA-IC with NSGA-II and MOGAIC with SPEA2 for simulation examples 2 and 3, respectively.

It can be seen from Figs. 5.7 and 5.8 that SPEA2 initially finds better solutions quickly as compared to NSGA-II but in the end cannot provide the best solutions. Furthermore, the search direction in both algorithms is clearly visible in Figs. 5.7 and 5.8. For example, with SPEA2, the search direction is from high-cost to low-cost regions (Figs. 5.7a and 5.8a), while maintaining several extreme solutions on each generations' pareto front. In contrast, the NSGA-II Pareto front moves toward the lowcost region without preserving each generations' extreme solutions. Instead, the entire Pareto front shifts as new solution sets are obtained. In other words, MOGA-IC with SPEA2 yields pareto fronts with wider spans, while MOGA-IC with NSGA-II distributes solutions in a more focused manner due to the different selection strategies used by NSGA-II and SPEA2.

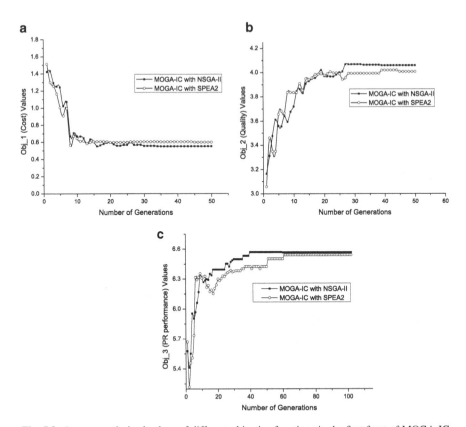

Fig. 5.8 Average optimized values of different objective functions in the first front of MOGA-IC with NSGA-II and SPEA2 for 100 generations. (**a**) Optimized avg. values of Obj_1 (cost) in NSGA-II and SPEA2. (**b**) Optimized avg. values of Obj_2 (quality) in NSGA-II and SPEA2 (**c**) Optimized avg. values of Obj_3 (PR performance) in NSGA-II and SPEA2

In MOGA-IC with NSGA-II, dominance ranking is used when forming the fronts of individuals, and these fronts are first used to populate the external set based on ranking, a strategy that allows a set of close-neighbor individuals in the same front to be included in the next generation. In contrast, MOGA-IC with SPEA2 selects individuals according to assigned fitness values based on Euclidean density information; so close-neighbor individuals are likely to be excluded in the next generation. The MOGA-IC with SPEA2, therefore, yields Pareto fronts with wider distributions of non-dominated solutions in contrast to MOGA-IC with NSGA-II, which is more focused when exploring the search space and generating Pareto solution sets.

Next, consider the simulation runtimes of both MOGAIC with NSGA-II and MOGA-IC with SPEA2, as shown in Table 5.8. From Table 5.8, we can see that MOGA-IC with NSGA-II runs much faster than does MOGA-IC with SPEA2 for the three simulation examples. The reason for this behavior is the time consumption

Table 5.8 Simulation runtimes of the three examples with MOGA-IC parameters

Simulation examples	m	R	N/E	G	P_c	P_m	Runtime (milliseconds)	
							NSGA-II	SPEA2
1	35	5	50	20	0.9	0.1	16.823	25.036
2	100	5	100	50	0.9	0.1	123.603	261.809
3	100	5	100	100	0.9	0.1	298.831	450.019

in the truncation approach. The time consumption of MOGA-IC with the NSGA-II truncation approach is much lower than that of MOGA-IC with SPEA2. This is mainly due to the superiority of the truncation approach in MOGA-IC with NSGA-II. The MOGA-IC with NSGA-II uses crowding distance as a truncation approach when the sizes of non-dominated solutions exceed the archive size. A crowding distance is the average distance of its neighbors along each of the objectives. The smaller is a solution's crowding distance, the more crowded is the area in which the solution may be located. NSGA-II only needs to sort all solutions on each objective, and so the time consumption of its truncation approach is not very sensitive to the number of non-dominated solutions. However, MOGA-IC with SPEA2 uses a truncation operator based on a nearest neighbor strategy, and the number of non-dominated solutions directly relates to the efficiency of the truncation approach in SPEA2. So we found that NSGA-II is the appropriate algorithm to develop MOGA-IC for the CP partner selection problem. Thus, the pCP can select any combination of CP partners from the pareto-optimal solution sets obtained from MOGA-IC based on NSGA-II.

5.3.4 Performance Comparison of MOGA-IC with MOGA-I in the CACM Model

In order to validate the proposed MOGA-IC model for CP partner selection in the CACM model, we develop another MOGA called MOGA-I based on NSGA-II that uses INIs for CP partner selection. We analyze the performances of the pCP that use both MOGA-IC and MOGA-I algorithms to make groups and to join various auctions in the CACM model. We assume that initially no collaborative information for other CPs is available to the pCP.

At the beginning of each auction, all providers including the pCP form several groups using MOGA-I and submit several group bids as single bids for a set of services to the auctioneer. The winner determination algorithm proposed in [98] is used to find the winners. Next, in the same auction with the same set of services, the winner determination algorithm is executed again, but this time ,the pCP uses the proposed MOGA-IC (others use a MOGA-I approach) to join the auctions and determine the winners. In our simulation, 1,000 auctions are generated for different user requirements. After each 100 auctions, we count the number of auctions won by the pCP using both algorithms. The experimental results are shown in Fig. 5.9.

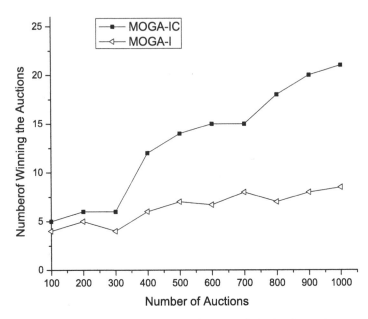

Fig. 5.9 Comparison of MOGA-IC with MOGA-I in terms of winning the auctions

It can be seen from Fig. 5.9 that using the MOGA-IC approach, pCP wins more auctions than it does using the MOGA-I approach. The reason is that the past collaborative performance values increase as the number of auctions increases, and as a result, the MOGA-IC finds a good combination of partners for pCP.

We also validate the performance of MOGA-IC to compare to that of MOGA-I in terms of conflicts reduction among the CP providers. We assume that conflicts may happen between providers P_{rj} and P_{xi} with the probability

$$p_{\text{conflicts}} = \begin{cases} \frac{1}{\delta \times e^{W_{rj,xi}}}, & \text{if } W_{rj,xi} \neq 0 \\ \frac{1}{\delta} & \text{otherwise, where } \delta \text{ is a constant} \end{cases} \quad i \neq j, r \neq x, \delta > 1 \quad (5.3)$$

We set $\delta = 20$ assuming that there is a 5% chance of conflicts between any two providers P_{rj} and P_{xi} if they have no past collaborative experience. Like the previous experiment, 1,000 auctions are generated. For each auction, when pCP uses both algorithms and forms groups, we count the total number of conflicts that may happen among the group members for various services using the probability conflicts. The experimental results are shown in Fig. 5.10. We can see from Fig. 5.10 that MOGA-IC can reduce a significant number of conflicts among providers as compared to the MOGA-I algorithm since it can utilize the PRI to choose partners along with the INI.

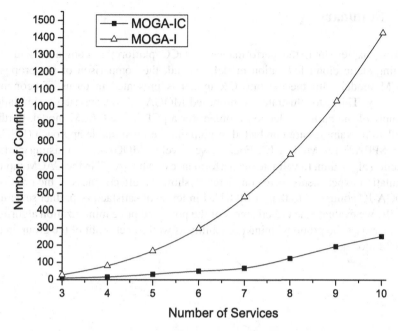

Fig. 5.10 Comparison of MOGA-IC with MOGA-I approach in terms of conflicts minimization

Table 5.9 MOGA-IC simulation results with various consumer service requirements

Simulation examples	m	R	N/E	G	P_c	P_m	Runtime (milliseconds) MOGA-IC with NSGA-II
1	100	6	80	100	0.9	0.1	214.882
2	100	8	100	200	0.9	0.1	480.187
3	100	10	100	200	0.9	0.1	500.689

5.3.5 Scalability Study of the MOGA-IC in Terms of Service Requirements

In the proposed CACM model, for each consumer requirement (R), there is a separate combinatorial auction. Based on the consumer requirements, a pCP runs MOGAIC to find appropriate CP partners to form groups. The MOGA-IC can support any number of consumer service requirements. To verify this, we conducted three simulations with MOGA-IC parameters as shown in Table 5.9.

From Table 5.9, we can see that the MOGA-IC's average run-time is mainly related to three parameters, population size (N/E), maximum genetic generation (G), and consumer service requirements (R). When they increase, the runtime will become longer. However, the run-time increases minimally with increasing R when the other parameters are fixed, such as in examples 2 and 3 in Table 5.9.

5.4 Summary

In this chapter, first, the performance of DCC platform is compared with the existing static cloud federation model. Second, the comparison of the proposed CACM model with the existing CA model is presented in terms of economic efficiency. Third, to illustrate the proposed MOGA-IC, we describe a simulation example of the partner selection problem for a pCP in the CACM model. Further simulation examples are conducted to pinpoint the most viable approach (NSGA-II or SPEA2) for MOGA-IC. Fourth, we develop MOGA-I, an existing partner selection algorithm, to validate the performance of MOGA-IC in the CACM model. Simulation experiments were conducted to show the effectiveness of the proposed MOGA-IC compared to that of MOGA-I in terms of satisfactory partner selection. Finally, we evaluate the effectiveness of the proposed price minimization algorithm that increases the group winning probability as well as net profit of the group in the CACM model.

Chapter 6
Closing Remarks

In today's world the emerging cloud computing offers a new computing model where resources such as computing power, storage, online applications and networking infrastructures can be shared as "services" over the Internet. However, the prevalent commercial CPs, operating in isolation (i.e. proprietary in nature), often face resource over-provisioning, degraded performance, and SLA violations (i.e. cloud service outages), thus incurring high operational costs and limiting the scope and scale of their services. Present trends in cloud service providers capabilities give rise to the interest in federating or collaborating clouds, hence allowing providers to revel on increased scale and reach than that is achievable individually.

Current research efforts in this context mainly focus on "Vertical Supply Chain Collaboration Model" in which cloud providers leverage cloud services from other cloud providers for seamless provisioning. Nevertheless, soon we can expect that hundreds of cloud providers will compete to offer services and thousands of users also compete to receive the services to run their complex heterogeneous applications on cloud computing environment. In this scenario, the existing federation/collaboration models are not applicable. In fact, while clouds are typically heterogeneous and dynamic, the existing federation models are designed for static environments where a-priori agreements among the parties are needed to establish the federation. Thus, a "Horizontal Collaboration Model" will emerge in which cloud providers (smaller, medium, and large) of complementary service requirements will collaborate dynamically to gain economies of scale and enlargements of their capabilities to meet QoS targets of heterogeneous cloud service requirements.

In this book, the technology for interconnection and inter-operation of cloud providers in the horizontal collaboration model is termed as "Dynamic Cloud Collaboration (DCC)". In this context, we have identified the fundamental research issues and challenges to address the core problems of when to collaborate, how to collaborate, whom to collaborate with and how to demonstrate dynamic collaborating applicability. We set forth our goals to address these key issues to achieve significant gains in cost effectiveness, performance, scalability, and coverage through the DCC framework.

M.M. Hassan and E.-N. Huh, *Dynamic Cloud Collaboration Platform:*   67
A Market-Oriented Approach, SpringerBriefs in Computer Science,
DOI 10.1007/978-1-4614-5146-4_6, © The Author(s) 2013

We present architectural framework and principles for the development of DCC. In particular, we propose a novel combinatorial auction-based cloud market model called CACM to enable a DCC platform among CPs which can fairly address the interoperability and scalability issues for cloud computing. The CACM model uses a new auction policy that allows CPs to dynamically collaborate with other partners and form groups and submit their group bids for a set of services as single bids. This policy can help to reduce collaboration cost as well as conflicts and negotiation time among CPs in DCC and therefore, creates more opportunities to win the auctions for the group. Furthermore, a new multi-objective optimization model of partner selection using the individual and past relationship information with collaboration cost optimization among partners is also proposed. An effective MOGA called MOGA-IC with NSGA-II is then developed to solve the model. The simulation results show that the MOGA-IC with NSGA-II is superior to the MOGA-IC with SPEA2 for solving the partner selection problem of CPs. Also in comparison with the existing MOGA-I approach; MOGA-IC with NSGA-II shows better performance results in CP partner selection in the CACM model.

The performance of the DCC platform is evaluated through extensive simulations. The results show that DCC platform effectively reduces the percentage of service rejection and improves the overall resource utilization as compared to the existing static collaboration approaches. Furthermore, the performance of the CACM model is justified as compared to the existing CA model in terms of economic efficiency. In addition, other simulation experiments show that MOGA-IC can select satisfactory partners in comparison with the existing partner selection approaches.

References

1. Software as a Service: Strategic Backgrounder. The Software Information Industry and Association. http://www.siia.net/estore/ssb-01.pdf. Accessed Feb 2001
2. Amazon S3 Availability Event. Amazon Service Health Dashboard. http://status.aws.amazon.com/s3-20080720.html. Accessed 20 July 2008
3. Amazon Elastic Compute Cloud (EC2). http://www.amazon.com/ec2/. Accessed March 2010
4. Google App Engine. http://appengine.google.com. Accessed Sept 2010
5. Windows azure platform. http://www.microsoft.com/azure/. Accessed Sept 2010
6. An, B., Lesser, V., Irwin, D., Zink, M.: Automated negotiation with decommitment for dynamic resource allocation in cloud computing. In: Proceedings of the 9th International Conference on Autonomous Agents and Multiagent Systems: Volume 1 - Volume 1, AAMAS '10, pp. 981–988 (2010)
7. Anandasivam, A., Buschek, S., Buyya, R.: A heuristic approach for capacity control in clouds. In: CEC '09: Proceedings of the 2009 IEEE Conference on Commerce and Enterprise Computing, pp. 90–97. IEEE Computer Society, Washington, DC (2009). doi: http://dx.doi.org/10.1109/CEC.2009.20
8. Armbrust, M., Fox, A., Griffith, R., Joseph, A.D., Katz, R.H., Konwinski, A., Lee, G., Patterson, D.A., Rabkin, A., Stoica, I., Zaharia, M.: Above the clouds: A berkeley view of cloud computing. Tech. Rep. UCB/EECS-2009-28, EECS Department, University of California, Berkeley (2009). http://www.eecs.berkeley.edu/Pubs/TechRpts/2009/EECS-2009-28.html
9. Bernstein, D., Ludvigson, E., Sankar, K., Diamond, S., Morrow, M.: Blueprint for the intercloud - protocols and formats for cloud computing interoperability. In: ICIW '09: Proceedings of the 2009 Fourth International Conference on Internet and Web Applications and Services, pp. 328–336. IEEE Computer Society, Washington, DC (2009). doi: http://dx.doi.org/10.1109/ICIW.2009.55
10. Boghosian, B., Coveney, P., Dong, S., Finn, L., Jha, S., Karniadakis, G., Karonis, N.: Nektar, spice and vortonics: using federated grids for large scale scientific applications. Cluster Comput. **10**(3), 351–364 (2007). doi:http://dx.doi.org/10.1007/s10586-007-0029-4
11. Boss, G., Malladi, P., Quan, D., Legregni, L., Hall, H.: IBM on cloud computing. IBM Technical Report, High Performance On-Demand Solutions (2007)
12. Bubendorfer, K.: Fine grained resource reservation in open grid economies. In: E-SCIENCE '06: Proceedings of the Second IEEE International Conference on e-Science and Grid Computing, p. 81. IEEE Computer Society, Washington, DC (2006). doi: http://dx.doi.org/10.1109/E-SCIENCE.2006.68
13. Bubendorfer, K., Thomson, W.: Resource management using untrusted auctioneers in a grid economy. In: E-SCIENCE '06: Proceedings of the Second IEEE International Conference on e-Science and Grid Computing, p. 74. IEEE Computer Society, Washington, DC (2006). doi: http://dx.doi.org/10.1109/E-SCIENCE.2006.115

M.M. Hassan and E.-N. Huh, *Dynamic Cloud Collaboration Platform:*
A Market-Oriented Approach, SpringerBriefs in Computer Science,
DOI 10.1007/978-1-4614-5146-4, © The Author(s) 2013

14. Buyya, R., Ranjan, R., Calheiros, R.: Intercloud: Utility-oriented federation of cloud computing environments for scaling of application services. In: Algorithms and Architectures for Parallel Processing. Lecture Notes in Computer Science, vol. 6081, pp. 13–31. Springer, Berlin (2010)

15. Calheiros, R.N., Ranjan, R., Beloglazov, A., Rose, C.A.F.D., Buyya, R.: CloudSim: a toolkit for modeling and simulation of cloud computing environments and evaluation of resource provisioning algorithms. J. Soft. Pract. Exp. **41**(1), 23–50 (2011)

16. Buyya, R., Yeo, C.S., Venugopal, S., Broberg, J., Brandic, I.: Cloud computing and emerging it platforms: Vision, hype, and reality for delivering computing as the 5th utility. Future Generat. Comput. Syst. **25**(6), 599–616 (2009). doi: http://dx.doi.org/10.1016/j.future.2008.12.001

17. Campbell, R., Gupta, I., Heath, M., Ko, S.Y., Kozuch, M., Kunze, M., Kwan, T., Lai, K., Lee, H.Y., Lyons, M., Milojicic, D., O'Hallaron, D., Soh, Y.C.: Open cirrustmcloud computing testbed: federated data centers for open source systems and services research. In: HotCloud'09: Proceedings of the 2009 conference on Hot topics in cloud computing, pp. 1–1. USENIX Association, Berkeley, CA (2009)

18. Celesti, A., Tusa, F., Villari, M., Puliafito, A.: How to enhance cloud architectures to enable cross-federation. In: IEEE 3rd International Conference on cloud Computing (CLOUD), pp. 337–345, Miami, FL, USA (2010)

19. Members of EGEE-II: An egee comparative study: Grids and clouds – evolution or revolution. Technical report, Enabling Grids for E-sciencE Project, June 2008. Electronic version available at https://edms.cern.ch/document/925013/

20. Chang, S.L., Wang, R.C., Wang, S.Y.: Applying fuzzy linguistic quantifier to select supply chain partners at different phases of product life cycle. Int. J. Prod. Econ. **100**(2), 348–359 (2006). http://ideas.repec.org/a/eee/proeco/v100y2006i2p348-359.html

21. Chang, Y.C., Li, C.S., Smith, J.R.: Searching dynamically bundled goods with pairwise relations. In: Proceedings of the 4th ACM conference on Electronic commerce, EC '03, pp. 135–143. ACM, New York, NY (2003). doi:http://doi.acm.org/10.1145/779928.779945. http://doi.acm.org/10.1145/779928.779945

22. Che, Z., Wang, H., Sha, D.: A multi-criterion interaction-oriented model with proportional rule for designing supply chain networks. Expert Syst. Appl. **33**(4), 1042–1053 (2007). DOI: 10.1016/j.eswa.2006.08.015. http://www.sciencedirect.com/science/article/B6V03-4KYY4GG-2/2/483f1f3b0da536fe558ae4ff5c0be975

23. Chen, Y.L., Cheng, L.C., Chuang, C.N.: A group recommendation system with consideration of interactions among group members. Expert Syst. Appl. **34**(3), 2082–2090 (2008). doi: http://dx.doi.org/10.1016/j.eswa.2007.02.008

24. Cheng, F., Ye, F., Yang, J.: Multi-objective optimization of collaborative manufacturing chain with time-sequence constraints. Int. J. Adv. Manuf. Tech. **40**, 1024–1032 (2009). http://dx. doi.org/10.1007/s00170-008-1388-6. 10.1007/s00170-008-1388-6

25. Chunyan, D., Yi, Y.: The method integration model for partner selection in the virtual enterprise. Int. Conf. Serv. Syst. Serv. Manag. **1**, 716–720 (2005). doi: http://doi.ieeecomputersociety.org/10.1109/ICSSSM.2005.1499570

26. Claburn, T.: Google news suffers outage. Information Week, http://www.informationweek. com/news/internet/google/showArticle.jhtml?articleID=220100659&subSection=Google. Accessed 22 Sept 2009

27. Clarke, E.H.: Multipart pricing of public goods. Publ. Choice **11**, 17–33 (1971)

28. Corne, D., Knowles, J., Oates, M.: The pareto envelope-based selection algorithm for multiobjective optimization. In: Parallel Problem Solving from Nature PPSN VI. Lecture Notes in Computer Science, vol. 1917, pp. 839–848. Springer, Berlin (2000)

29. Coursey, D.: Google outage: If this is our future, it looks bad. PC World, http://www.gartner. com/it/page.jsp?id=871113. Accessed 01 Sept 2009

30. Cowan, R., Jonard, N., Zimmermann, J.B.: Bilateral collaboration and the emergence of innovation networks. Manag. Sci. **53**(7), 1051–1067 (2007). doi: http://dx.doi.org/10.1287/mnsc.1060.0618

31. Das, A., Grosu, D.: Combinatorial auction-based protocols for resource allocation in grids. In: IPDPS '05: Proceedings of the 19th IEEE International Parallel and Distributed Processing Symposium (IPDPS'05) - Workshop 13, p. 251.1. IEEE Computer Society, Washington, DC (2005). doi:http://dx.doi.org/10.1109/IPDPS.2005.140

32. Deb, K., Agrawal, S., Pratap, A., Meyarivan, T.: A fast elitist non-dominated sorting genetic algorithm for multi-objective optimisation: Nsga-ii. In: PPSN VI: Proceedings of the 6th International Conference on Parallel Problem Solving from Nature, pp. 849–858. Springer, London (2000)

33. Ding, J.F., Liang, G.S.: Using fuzzy mcdm to select partners of strategic alliances for liner shipping. Int. J. Informat. Comput. Sci. **173**(1–3), 197–225 (2005)

34. Dodda, R.T., Smith, C., van Moorsel, A.: An architecture for cross-cloud system management. Contemporary computing, communications in computer and information science, vol. 40. Springer, Berlin (2009). ISBN 978-3-642-03546-3

35. Dubey, A., Mohiuddin, J., Baijal, A., Rangaswami, M.: Enterprise software customer survey. Sand Hill Group, McKinsey and Company. http://www.interop.com/downloads/mckinsey_interop_survey.pdf (2008)

36. Elmroth, E., Larsson, L.: Interfaces for placement, migration, and monitoring of virtual machines in federated clouds. In: GCC '09: Proceedings of the 2009 Eighth International Conference on Grid and Cooperative Computing, pp. 253–260. IEEE Computer Society, Washington, DC, USA (2009). doi:http://dx.doi.org/10.1109/GCC.2009.36

37. Emden, Z., Calantone, R., Droge, C.: Collaborating for new product development: selecting the partner with maximum potential to create value. J. Prod. Innovat. Manag. **23**(4), 330–341 (2006)

38. Famuyiwa, O., Monplaisir, L., Nepal, B.: An integrated fuzzy-goal-programming-based framework for selecting suppliers in strategic alliance formation. Int. J. Prod. Econ. **113**(2), 862–875 (2008)

39. Fan, Z.P., Bo, F.: A multiple attributes decision making method using individual and collaborative attribute data in a fuzzy environment. Inform. Sci. **179**(20), 3603–3618 (2009)

40. Fan, Z.P., Feng, B., Jiang, Z.Z., Fu, N.: A method for member selection of R&D teams using the individual and collaborative information. Expert Syst. Appl. **36**(4), 8313–8323 (2009). doi:http://dx.doi.org/10.1016/j.eswa.2008.10.020

41. Farahani, R.Z., Elahipanah, M.: A genetic algorithm to optimize the total cost and service level for just-in-time distribution in a supply chain. Int. J. Prod. Econ. **111**(2), 229–243 (2008)

42. Feng, B., Fan, Z.P., Ma, J.: A method for partner selection of codevelopment alliances using individual and collaborative utilities. Int. J. Prod. Econ. **124**(1), 159–170 (2010)

43. Fischer, M., Jahn, H., Teich, T.: Optimizing the selection of partners in production networks. Int. J. Robot. Comput. Integr. Manuf. **20**(6), 593–601 (2004)

44. Fuqing, Z., Yi, H., Dongmei, Y.: A multi-objective optimization model of the partner selection problem in a virtual enterprise and its solution with genetic algorithms. Int. J. Adv. Manuf. Tech. **28**, 1246–1253 (2006). http://dx.doi.org/10.1007/s00170-004-2461-4. 10.1007/s00170-004-2461-4

45. Gartner, I.: Cloud application infrastructure technologies need seven years to mature. http://www.gartner.com/it/page.jsp?id=871113. Accessed Feb 2009

46. Geelan, J.: Twenty-one experts define cloud computing. Retrieved Sept 2010 from http://cloudcomputing.sys-con.com/node/612375/print. Accessed Aug 2008

47. Gens, F.: IDC's new IT cloud services forecast: 2009–2013. http://blogs.idc.com/ie/?p=543. Accessed Oct 2009

48. Goiri, I., Guitart, J., Torres, J.: Characterizing cloud federation for enhancing providers' profit. In: IEEE 3rd International Conference on Cloud Computing, pp. 123–130, Miami, FL, USA (2010)

49. Gomes, E.R., Vo, Q.B., Kowalczyk, R.: Pure exchange markets for resource sharing in federated clouds. Concurrency Comput. Pract. Exp. **24**(9), 977–991 John Wiley and Sons Ltd. Chichester, UK (2012)

50. Hajidimitriou, Y.A., Georgiou, A.C.: A goal programming model for partner selection decisions in international joint ventures. Eur. J. Oper. Res. **138**(3), 649–662 (2002)

51. Hoffa, C., Mehta, G., Freeman, T., Deelman, E., Keahey, K., Berriman, B., Good, J.: On the use of cloud computing for scientific workflows. In: ESCIENCE '08: Proceedings of the 2008 Fourth IEEE International Conference on eScience, pp. 640–645. IEEE Computer Society, Washington, DC, USA (2008). doi:http://dx.doi.org/10.1109/eScience.2008.167

52. Huang, B., Gao, C., Chen, L.: Study on partner selection for a virtual enterprise based on vague sets. In: CCIE'10: Proceedings of the 2010 International Conference on Computing, Control and Industrial Engineering, 1, pp. 110–113, Wuhan, China (2010)

53. Huang, X.G., Wong, Y.S., Wang, J.: A two-stage manufacturing partner selection framework for virtual enterprises. Int. J. Comput. Integr. Manuf. **17**(4), 294–304 (2004). http://www.informaworld.com/10.1080/09511920310001654292

54. IBM: Cloud computing. https://www.ibm.com/developerworks/cloud/. Accessed Sept 2010

55. Ip, W.H., Huang, M., Yung, K.L., Wang, D.: Genetic algorithm solution for a risk-based partner selection problem in a virtual enterprise. Comput. Oper. Res. **30**(2), 213–231 (2003). doi:http://dx.doi.org/10.1016/S0305-0548(01)00092-2

56. Jarimo, T., Salo, A.: Multicriteria partner selection in virtual organizations with transportation costs and other network interdependencies. Trans. Syst. Man Cybern. C **39**(1), 124–129 (2009)

57. Kaya, M.: Mogamod: Multi-objective genetic algorithm for motif discovery. Expert Syst. Appl. **36**(2, Part 1), 1039–1047 (2009). DOI: 10.1016/j.eswa.2007.11.008. http://www.sciencedirect.com/science/article/B6V03-4R53W74-M/2/e6cf5fe5acf73b555a60d248f6f59155

58. Klems, M.: Merrill Lynch Estimates "cloud computing" to be $ 100 Billion Market. http://web2.sys-con.com/node/604936. Accessed 2012

59. Ko, C.S., Kim, T., Hwang, H.: External partner selection using tabu search heuristics in distributed manufacturing. Int. J. Prod. Res. **39**(17), 3959–3974 (2001)

60. Konak, A., Coit, D., Smith, A.: Multi-objective optimization using genetic algorithms: A tutorial. Reliab. Eng. Syst. Saf. **91**, 992–1007 (2006)

61. Li, W., Ping, L.: Trust model to enhance security and interoperability of cloud environment. In: CloudCom '09: Proceedings of the 1st International Conference on Cloud Computing, pp. 69–79. Springer, Berlin (2009). doi:http://dx.doi.org/10.1007/978-3-642-10665-1_7

62. Lublin, U., Feitelson, D.G.: The workload on parallel supercomputers: modeling the characteristics of rigid jobs. J. Parallel Distr. Comput. **63**(11), 1105–1122 (2003). doi: http://dx.doi.org/10.1016/S0743-7315(03)00108-4

63. Maximilien, E.M., Ranabahu, A., Engehausen, R., Anderson, L.: Ibm altocumulus: a cross-cloud middleware and platform. In: OOPSLA '09: Proceeding of the 24th ACM SIGPLAN Conference Companion on Object Oriented Programming Systems Languages and Applications, pp. 805–806. ACM, New York (2009). doi: http://doi.acm.org/10.1145/1639950.1640024

64. McLaughlin, K.: Netsuite back online after cloud apps outage. CRN, http://www.crn.com/news/applications-os/224600702/netsuite-back-online-after-cloud-apps-outage.htm; jsessionid=nUYkEwwMtpscjA+khR-7iw**.ecappj02. Accessed 27 April 2010

65. Miller, R.: Brief power outage for Amazon data center. Data Center Knowledge. http://www.datacenterknowledge.com/archives/2009/12/10/power-outage-for-amazon-data-center/. Accessed 10 Dec 2009

66. Nepal, S., Zic, J.: A conflict neighbouring negotiation algorithm for resource services in dynamic collaborations. In: SCC '08: Proceedings of the 2008 IEEE International Conference on Services Computing, pp. 283–290. IEEE Computer Society, Washington, DC (2008). doi: http://dx.doi.org/10.1109/SCC.2008.18

67. Nepal, S., Zic, J., Chan, J.: A distributed approach for negotiating resource contributions in dynamic collaboration. In: Proceedings of the Eighth International Conference on Parallel and Distributed Computing, Applications and Technologies, pp. 82–86. IEEE Computer Society, Washington, DC (2007). doi:http://dx.doi.org/10.1109/PDCAT.2007.1

68. OGF: Open cloud computing interface working group. http://www.occi-wg.org/. Accessed 2012
69. OpenQRM: The next generation, open-source data-center management platform. http://www.openqrm.com/. Accessed Sept 2010
70. Paul, I.: Google suffers another service outage. PC World. http://www.pcworld.com/article/165046/google_suffers_another_service_outage.html. Accessed 18 May 2009
71. Petersen, S.A., Matskin, M.: Agent interaction protocols for the selection of partners for virtual enterprises. In: CEEMAS'03: Proceedings of the 3rd Central and Eastern European conference on Multi-agent systems, pp. 606–615. Springer, Berlin (2003)
72. Ranjan, R., Buyya, R.: Decentralized overlay for federation of enterprise clouds. Handbook of Research on Scalable Computing Technologies. ISBN: 978-1-60566-661-7, IGI Global (2009)
73. Raphae, J.: Google says outage caused by traffic routing error. PC World. http://www.pcworld.com/article/164904/google_says_outage_caused_by_traffic_routing_error.html?tk=rel_news. Accessed 15 May 2009
74. Rochwerger, B., Breitgand, D., Levy, E., Galis, A., Nagin, K., Llorente, I.M., Montero, R., Wolfsthal, Y., Elmroth, E., Cáceres, J., Ben-Yehuda, M., Emmerich, W., Galán, F.: The reservoir model and architecture for open federated cloud computing. IBM J. Res. Dev. 53(4), 535–545 (2009)
75. Saen, R.F.: Suppliers selection in the presence of both cardinal and ordinal data. Eur. J. Oper. Res. 183(2), 741–747 (2007)
76. Salesforce.com: Application development with force.coms cloud computing platform. http://www.salesforce.com/platform/. Accessed Sept 2010
77. Sarker, R., Liang, K.H., Newton, C.: A new multiobjective evolutionary algorithm. Eur. J. Oper. Res. 140(1), 12–23 (2002)
78. Schofield, J.: Google angles for business users with platform as a service. The Guardian. http://www.guardian.co.uk/technology/2008/apr/17/google.software (2008). Accessed 2012
79. Shantanu, B.: Design of Iterative Mechanisms for Combinatorial Auctions and Exchanges. Ph.D Thesis, Computer Science and Automation Indian Institute of Science Bangalore (2004)
80. Sobolewski, M., Kolonay, R.M., Sobolewski, M., Kolonay, R.M.: Federated grid computing with interactive service-oriented programming. Int. J. Concurrent Eng. Res. Appl. 14, 55–66 (2006)
81. Sotomayor, B., Montero, R.S., Llorente, I.M., Foster, I.: Resource leasing and the art of suspending virtual machines. In: 11th IEEE International Conference on High Performance Computing and Communications, pp. 59–68, Seoul, South Korea (2009)
82. Stanley, M.: Technology trends. http://www.morganstanley.com. Accessed June 2008
83. Sun: Sun Microsystems' open cloud platform. http://www.oracle.com/us/sun/index.html. Accessed Sept 2010
84. Suznjevic, M., Matijasevic, M., Dobrijevic, O.: Action specific massive multiplayer online role playing games traffic analysis: case study of world of warcraft. In: NetGames'08: Proceedings of the 7th ACM SIGCOMM Workshop on Network and System Support for Games, pp. 106–107. ACM, New York, USA (2008)
85. Suzuki, K., Yokoo, M.: Secure generalized vickrey auction using homomorphic encryption. In: Financial Cryptography. Lecture Notes in Computer Science, vol. 2742, pp. 239–249. Springer, Berlin (2003)
86. Tubanos, A.: Flexiscale suffers 18-hour outage. http://www.thewhir.com/web-hosting-news/103108_FlexiScale_Suffers_18_Hour_Outage. Accessed 31 Oct 2008
87. Van Hoesel, S., Muller, R.: Optimization in electronic markets: examples in combinatorial auctions. Netnomics 3(1), 23–33 (2001)
88. Varvello, M., Voelker, G.M.: Second Life: a social network of humans and bots. In: Nosdav'2010, 20th International Workshop on Network and Operating Systems Support for Digital Audio and Video, 2–4 June 2010, Amsterdam, The Netherlands (2010). doi:10.1145/1806565.1806570

89. Vickrey, W.: Counterspeculation, auctions, and competitive sealed tenders. J. Finance **16**, 8–37 (1961)
90. Wang, Z.J., Xu, X.F., Zhan, D.C.: Genetic algorithms for collaboration cost optimization-oriented partner selection in virtual enterprises. Int. J. Prod. Res. **47**(4), 859–881 (2009)
91. Williams, A.: Top 5 cloud outages of the past two years: Lessons Learned. http://www.readwriteweb.com/cloud/2010/02/top-5-cloud-outages-of-the-pas.php. Accessed Feb 2010
92. Wu, N., Su, P.: Selection of partners in virtual enterprise paradigm. Robot. Comput. Integr. Manuf. **21**(2), 119–131 (2005). DOI: 10.1016/j.rcim.2004.05.006. http://www.sciencedirect.com/science/article/B6V4P-4D4D1GP-2/2/ef5b5f277e3dc17aadba73be632c2771
93. Yang, T., Hung, C.C.: Multiple-attribute decision making methods for plant layout design problem. Robot. Comput. Integr. Manuf. **23**(1), 126–137 (2007). doi: http://dx.doi.org/10.1016/j.rcim.2005.12.002
94. Ye, F., Li, Y.N.: Group multi-attribute decision model to partner selection in the formation of virtual enterprise under incomplete information. Expert Syst. Appl. **36**(5), 9350–9357 (2009). doi:http://dx.doi.org/10.1016/j.eswa.2009.01.015
95. Yeh, W.C., Chuang, M.C.: Using multi-objective genetic algorithm for partner selection in green supply chain problems. Expert. Syst. Appl. **38**(4), 4244–4253 (2011)
96. Yen, G., Lu, H.: Dynamic multiobjective evolutionary algorithm: adaptive cell-based rank and density estimation. IEEE Trans. Evol. Comput. **7**(3), 253–274 (2003)
97. Yeo, C.S., Venugopal, S., Chu, X., Buyya, R.: Autonomic metered pricing for a utility computing service. Future Generat. Comput. Syst. **26**(8), 1368–1380 (2010). doi: http://dx.doi.org/10.1016/j.future.2009.05.024
98. Yokoo, M., Suzuki, K.: Secure multi-agent dynamic programming based on homomorphic encryption and its application to combinatorial auctions. In: AAMAS '02: Proceedings of the first international joint conference on Autonomous agents and multiagent systems, pp. 112–119. ACM, New York (2002). doi:http://doi.acm.org/10.1145/544741.544770
99. Youseff, L., Butrico, M., Da Silva, D.: Toward a unified ontology of cloud computing. In: Grid Computing Environments Workshop, 2008. GCE '08, pp. 1–10 (2008). doi:10.1109/GCE.2008.4738443. http://dx.doi.org/10.1109/GCE.2008.4738443
100. Zeng, Z., Li, Y., Li, S., Zhu, W.: A new algorithm for partner selection in virtual enterprise. In: PDCAT '05: Proceedings of the Sixth International Conference on Parallel and Distributed Computing Applications and Technologies, pp. 884–886. IEEE Computer Society, Washington, DC (2005). doi:http://dx.doi.org/10.1109/PDCAT.2005.22
101. Zhong, Y., Jian, L., Zijun, W.: An integrated optimization algorithm of GA and ACA-based approaches for modeling virtual enterprise partner selection. SIGMIS Database **40**(2), 37–56 (2009)
102. Zitzler, E., Laumanns, M., Thiele, L.: SPEA2: Improving the strength pareto evolutionary algorithm for multiobjective optimization. In: Evolutionary Methods for Design Optimization and Control with Applications to Industrial Problems, pp. 95–100. International Center for Numerical Methods in Engineering, Athens, Greece (2001)
103. Zizler, E., Thiele, L.: Multiobjective evolutionary algorithms: a comparative case study and the strength Pareto approach. IEEE Trans. Evol. Comput. **3**(4), 257–271 (1999)